Praise for *Things That Art*

"At once a meditation and a manifesto, *Things That Art* shows the rich, strange, and arbitrary ways that meaning is made by queer things and outcasts of all kinds. Jain moves them from the margins to a place of exuberant centrality."
– **Brian Selznick and David Serlin, creators of *Baby Monkey, Private Eye***

"Wild and creative scribbles for sparking fresh ideas."
– **Lauren Baker, London-based visual artist**

"*Things That Art* is deliciously subversive of normative culture – readers are invited to savor the strangeness of language and history via Jain's fiendishly clever illustrations."
– **Joseph Masco, professor of anthropology at the University of Chicago**

"A fascinating book, with an unusual proposition. The three contributors weave a patchwork of visual, theoretical, and philosophical information that can be viewed as one composition or as interesting details."
– **Pia MYrvoLD, visual artist and designer**

LOCHLANN JAIN

THINGS THAT ART

a graphic menagerie of enchanting curiosity

UNIVERSITY OF TORONTO PRESS
Toronto Buffalo London

Toronto Buffalo London
utorontopress.com
Printed in Canada

Library and Archives Canada Cataloguing in Publication

Title: Things that art : a graphic menagerie of enchanting curiosity / Lochlann Jain.
Names: Jain, Lochlann, author, artist.
Description: Series statement: ethnoGRAPHIC | Includes bibliographical references.
Identifiers: Canadiana 20190128585 | ISBN 9781487570552 (hardcover)
Subjects: LCSH: Art and anthropology.
Classification: LCC NC139.J35 T55 2019 | DDC 741.973 – dc23

North America
5201 Dufferin Street
North York, Ontario, Canada, M3H 5T8

2250 Military Road
Tonawanda, New York, USA, 14150
ORDERS PHONE: 1-800-565-9523
ORDERS FAX: 1-800-221-9985
ORDERS E-MAIL: utpbooks@utpress.utoronto.ca

UK, Ireland, and continental Europe
NBN International
Estover Road, Plymouth, PL6 7PY, UK
ORDERS PHONE: 44 (0) 1752 202301
ORDERS FAX: 44 (0) 1752 202333
ORDERS E-MAIL: enquiries@nbninternational.com

University of Toronto Press acknowledges the financial assistance to its publishing program of the Canada Council for the Arts and the Ontario Arts Council, an agency of the Government of Ontario.

Canada Council for the Arts | Conseil des Arts du Canada

ONTARIO ARTS COUNCIL
CONSEIL DES ARTS DE L'ONTARIO
an Ontario government agency
un organisme du gouvernement de l'Ontario

Funded by the Government of Canada | Financé par le gouvernement du Canada | Canada

FSC www.fsc.org MIX Paper from responsible sources FSC® C016245

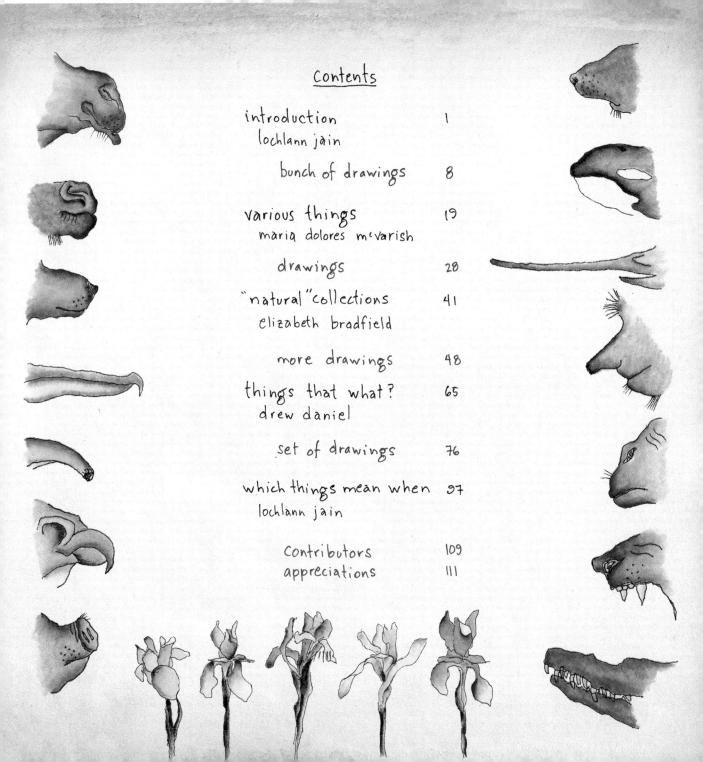

Contents

things in a pen

e   pig   ink   drawing

Løchlann

Introduction

LOCHLANN JAIN

All hail the nose. Not just any nose, mind you, but the flawless, magnetic Caesar of a colleague, which, one afternoon, yanked my attention from the lobs and volleys of a committee meeting. Awestruck and unable to help myself, I jotted its likeness in my agenda.

This induced a reverie on "kinds of noses" and I began ruminating on the category, casting for more specimens. A bulbous ski slope transpired, which I jokingly labeled "my sister's." This comforting paean to kinship momentarily eased my alienation from the committee's discussion. In short order, the Important Matters under debate gently receded as new matters gained import. My unleashed hand crowded the page with the materially comic (a clown's foam ball), the conceptual (a drawing of a drawing of a nose ripped in two: out of joint), the uncanny (equine, porcine), and the disparaged (racialized). This last nose evoked the centuries of looked-down, turned-up ivory-tower noses that had unwittingly initiated the mini-gallery. "Standard" nose, I sneeze at you!

One day while drawing a group of seven irises from his garden, John Berger mused, "We who draw do so not only to make something visible to others, but also to accompany something invisible to its incalculable destination."[1] I love this vision of the artist escorting a beloved companion into an interdependent existence. How delightful to share a journey without a foreseeable end, and to cultivate dialogue and friendship through that process. *Things That Art* proposes that hand-rendering new content for insertion into a traditional form of visual display offers access to the shadowy internalized images that serve as shaky bedrocks and clammy wellsprings for everyday assumptions. How do our own irises glimpse the bulb in bulbous, how do our schnozzles distinguish spring blossoms?

In a life drawing class, a professor will demand that students look closely – an idea of how a hand *should* look will only lead neophytes astray when they attempt to reproduce the odd forms of the *actual* knuckle before them. By prioritizing the preconceived notion over the original or artistic vision, *Things That Art* offers something different. The design I settled on, columns and rows of labeled boxes with empty space between, practically demands a standardized version of each thing – visual descriptions learned not by close examination of an actual dragonfly or tibia, but by recalling a diagram, an illustration, or a stereotype. A unique line may muscle in, dispensing a charming variation on remembered shapes, but the point is to materialize a memory rather than to faithfully represent a lemon, skull, and goblet behind an easel.

Needless to say, the initial scribbles nosed toward a full-fledged project. Stamp-sized drawings were done in pen, from memory, without judgment, on 4 × 6-inch watercolor pads. These self-imposed guidelines stymied any imposter complex (Damn it Jain! You're a scholar, not a sketcher!) and limited my scope for catastrophe. I was free to simply draw from memory likenesses of objects I had only occasionally or never seen: a salamander, a shadow of doubt, or a pirate. Like the best imaginary concoctions, many of the drawings became friends.

In my bailiwick as an anthropologist I study people and their interactions with stuff: cars, laws, and viruses, for instance. I unpack the very strange histories of familiar, taken-for-granted facts of life. As an artist, I create things: cartoons, prints, and paintings. With no discernible purpose other than attracting an eye or eliciting a chuckle, they can offer a way to proliferate and process questions in ways not available through traditional scholarly methods that trade in explanation. Initially intrigued by the sorts of juxtapositions that emerged unbidden from my pen, I also came to see that the graphic menagerie emerging in my growing collection of cards not only gave me a warm feeling of popularity among my new posse of paper mates, but enabled me to reimagine engagements with age-old philosophical questions about the relations among word and image, category and individual, hand-drawn and mass-produced lines, and label and collection. A sketched album could invert stereotypes and generate queer ways of tracking scents, releasing doves with diverse messages to explore new flight paths between pigeonholes.

The form of my drawings will be familiar, recalling the picture postcard, the botanical color plate, the baseball trading card. Zoos, art galleries, and museums adopt a parallel scaffold. Each framed or caged thing harnesses the same design principle as the lowly flashcard and child's alphabet book. Such displays patently show *and* tell in a mutually illustrative circuit. At the tech museum, a displayed Macintosh SE computer, 1989, from Silicon Valley will be labeled "Macintosh SE computer, 1989, Silicon

Valley." We see, we recognize, we read, we know, we reiterate. This is the stuff of Western education assimilated by miniature humans from 8:15 am to 3:25 pm five days a week, for years pasted to seats attached to desks assembled in columns and rows, feet flat on a classroom's linoleum floor, quiet, hands to themselves, writing or occasionally raised (but no poking! no doodling!), learning the arts of docility, looking at words and pictures so as to reproduce the words if not the pictures in future exams that may or may not bear on life chances. Most of us have been there; some of us are still recovering.

The seamlessness of word and image in this circuit obscures the hierarchical interests that encourage sifting, sorting, and collecting in some ways over others. These produce an order of things in the world that is handed over as fully formed fact. As someone who has spent altogether too much time around dictionaries, encyclopedias, and museums, it was no accident that this idiom invaded my scrawls and, for this moment, yours as well.

Nothing if not useful, the thing+label genre does tender a fragile thread to the sentient world. Consider your last trip to the zoo. Visitors may disagree about whether polar bears should be in cages, but we all accede that what prowls behind the glass is from up north and that it is bigger and hairier than the snake in the next pavilion. This diaphanous concept of *polar bear*, gleaned between the kiddie train and a frayed nerve, can't compare to what the Inuit know. Yet it forms the basis of a shared understanding.

Leaving the zoo for the museum, one might come upon a plinthed assemblage presented with a brass plate: *Hippopotamus amphibius*. Never mind that the hide of the original hippo was peeled off its fleshy owner and stuffed with sawdust several thousand miles later. Never mind the virtual impossibility of imagining the evacuated life force – the conversational hippo grunts with chums while munching the sweet grass of the Okavango Delta. The series of unpleasant encounters that led to the appearance of this solitary, spiritless aggregate hides behind the painted weeds and crumbling plaster of the diorama in London or New York or Rome.

Giving my pen over to the spontaneity of the form sometimes yielded groupings that I didn't fully understand myself, in part because of language's arbitrary quality, which linguists and grand theorists have attempted to overcome somewhat fruitlessly (with all due respect) since at least the beginnings of philosophy. While naming and organizing the world and all that's in it has been something of an obsession among men with pens, those who have been squashed into categories for convenience or out of confusion do add a unique perspective.

Just ask the platypus, a cutie with whom I strongly identify. First disemboweled and sent to London in 1798 by an Australian governor, the poor soul's dapper fur attire and egg-laying ritual wrought consternation in the metropole, sparking an 85-year-long battle about whether to slot this curio into the animal kingdom at all. That "first" platypus soon became an exemplar specimen used to judge subsequent platypodes; it still resides in London's Natural History Museum, in a drawer, with a label clinging to its toe.

Swedish botanist Carl Linnaeus, inventor of a binomial classification system from which no animal or plant could escape, found a way to account for these un-naturals of history with two special taxa. Into *monstrosus* he placed human savages, noble and otherwise. *Paradoxa* contained the phoenix, dragon, and manticore. Even the penniless old pelican took up temporary residence in the *paradoxa* halfway house, falsely accused of feeding blood to her young through a self-stabbing ritual. Others in that category, orthrus or abaia, might have escaped the dime museum or freak show, but beware of the hedged existential bets of binomial classification and its awkwardness over those who are more than one and less than two. The term *Ornithorhynchus paradoxus*, used by Charles Darwin himself, swathed the bewildered platypus in a crisis of identity for nearly two centuries. While it can be a blow to one's cool, every hybrid (one hopes) grasps that the flaw rests not within them but the accounting system.

Categories have their uses. They enable concepts and organize perception, and in so doing, they constitute those who devise them, those who are ensnared by them, and the worlds in which they move together. They dispense opportunity for some and the opposite for anyone not fully invested in their proper slot. But if categories need us and we need them, what scope exists for revision? When things falter, do we fail – in our identity performances, the value of our social contributions, our modes of being? Maybe. In my view it's a worthy risk. Adding "dis" onto "order" will enable new hodgepodges, hocus-pocus, and hanky-panky to burst from the ruins.

The very first sketch of that exquisite professorial nose – the very incarnation of the nose that gets to know – made apparent that living, lying, consensual paradoxes could be drawn together, drawn out, drawn into being, and drawn nearer. If nothing else, prying word from image opens a good craic.

Hopefully this aerated, kaleidoscopic, and woolly graphic menagerie will inspire you to diverge from your rank and file, if only for an instant. All hail a poke toward your neighbor! Or perhaps you'd prefer to accompany a squiggle resembling your colleague's body part to its incalculable destination. Discretion is advised (sort of).

Note

1 John Berger, *Bento's Sketchbook* (London: Verso, 2015), 9.

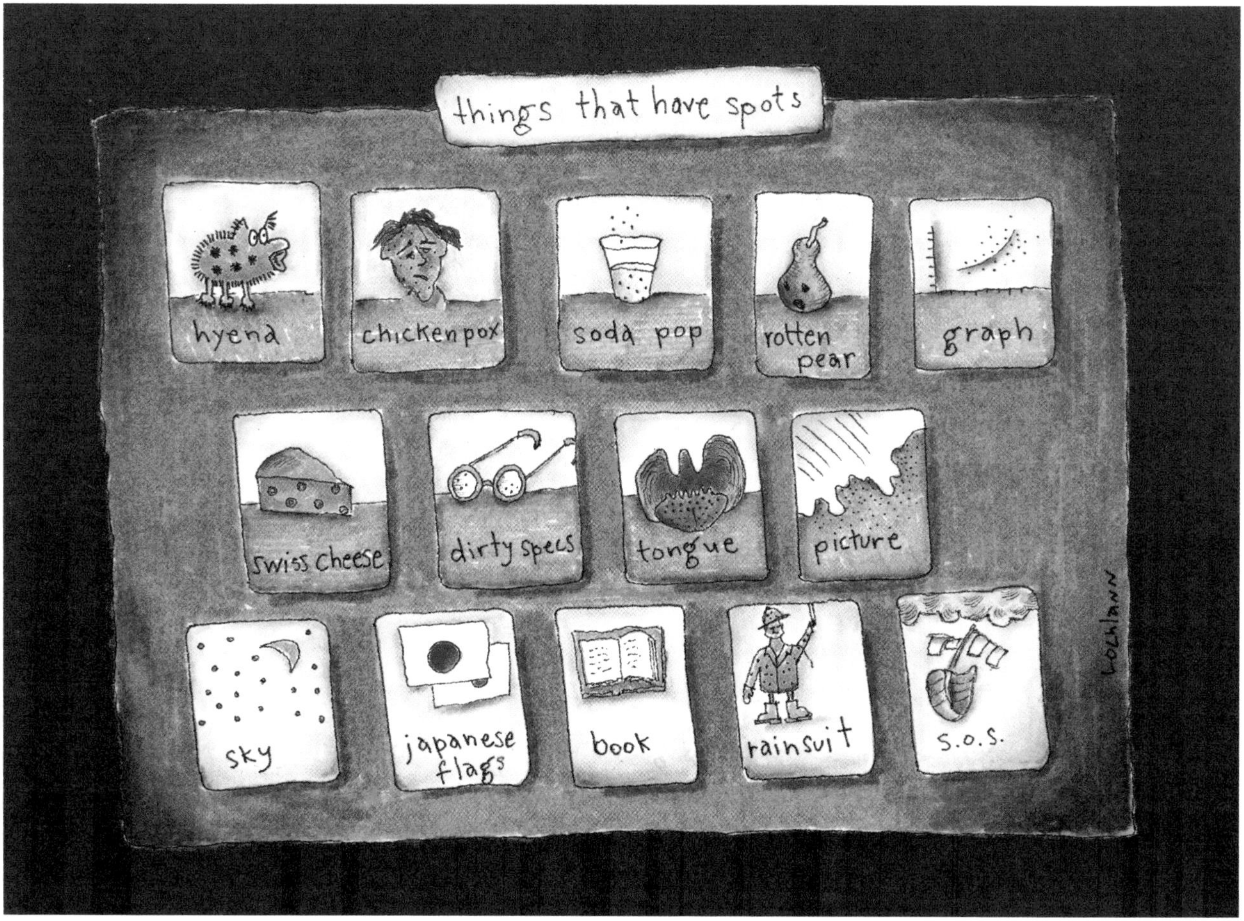
things that have spots
hyena
chickenpox
soda pop
rotten pear
graph
swiss cheese
dirty specs
tongue
picture
sky
japanese flags
book
rainsuit
s.o.s.
LOCHLANN

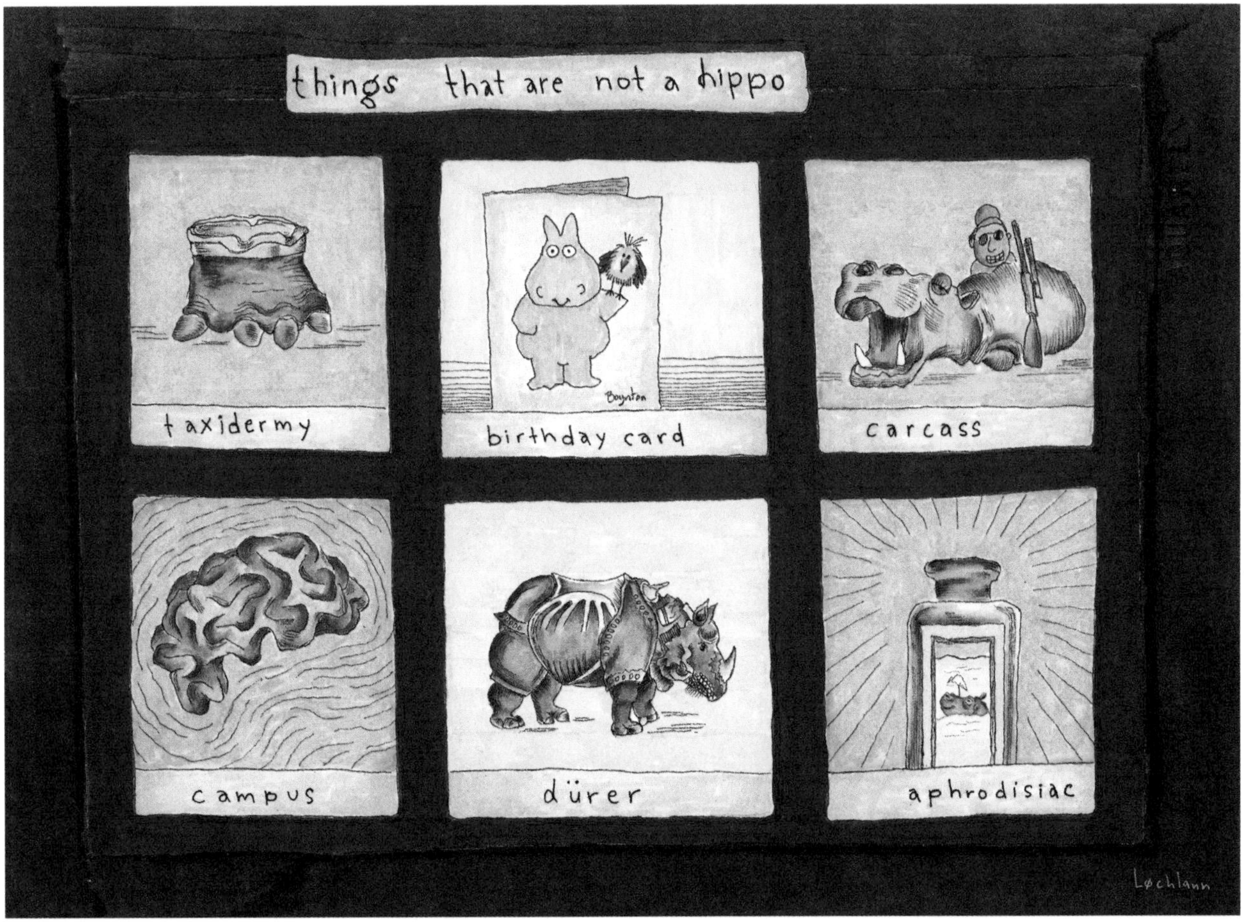
things that are not a hippo
taxidermy
birthday card
Boynton
carcass
campus
dürer
aphrodisiac
Løchlann

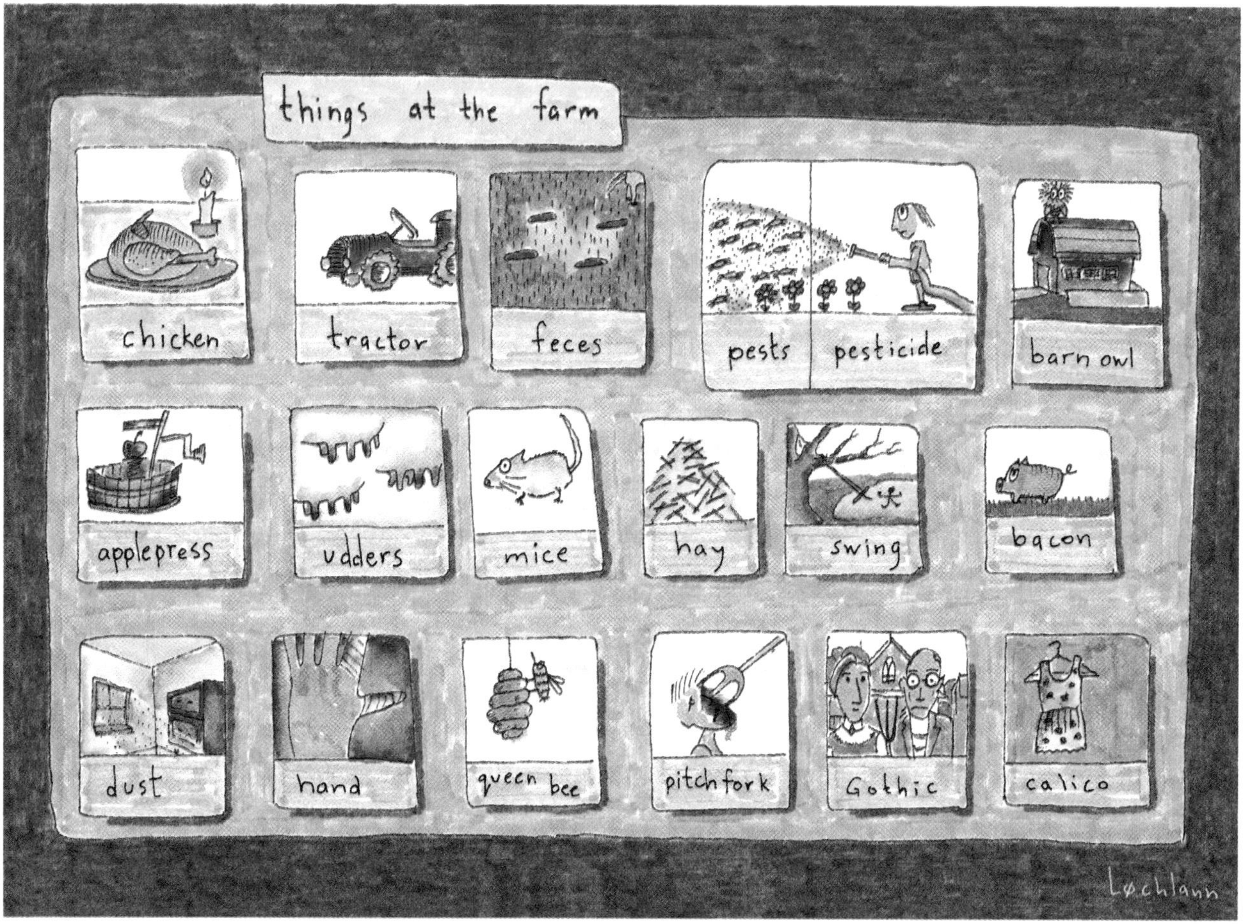
things at the farm
chicken
tractor
feces
pests
pesticide
barn owl
applepress
udders
mice
hay
swing
bacon
dust
hand
queen bee
pitchfork
Gothic
calico
Løchlann

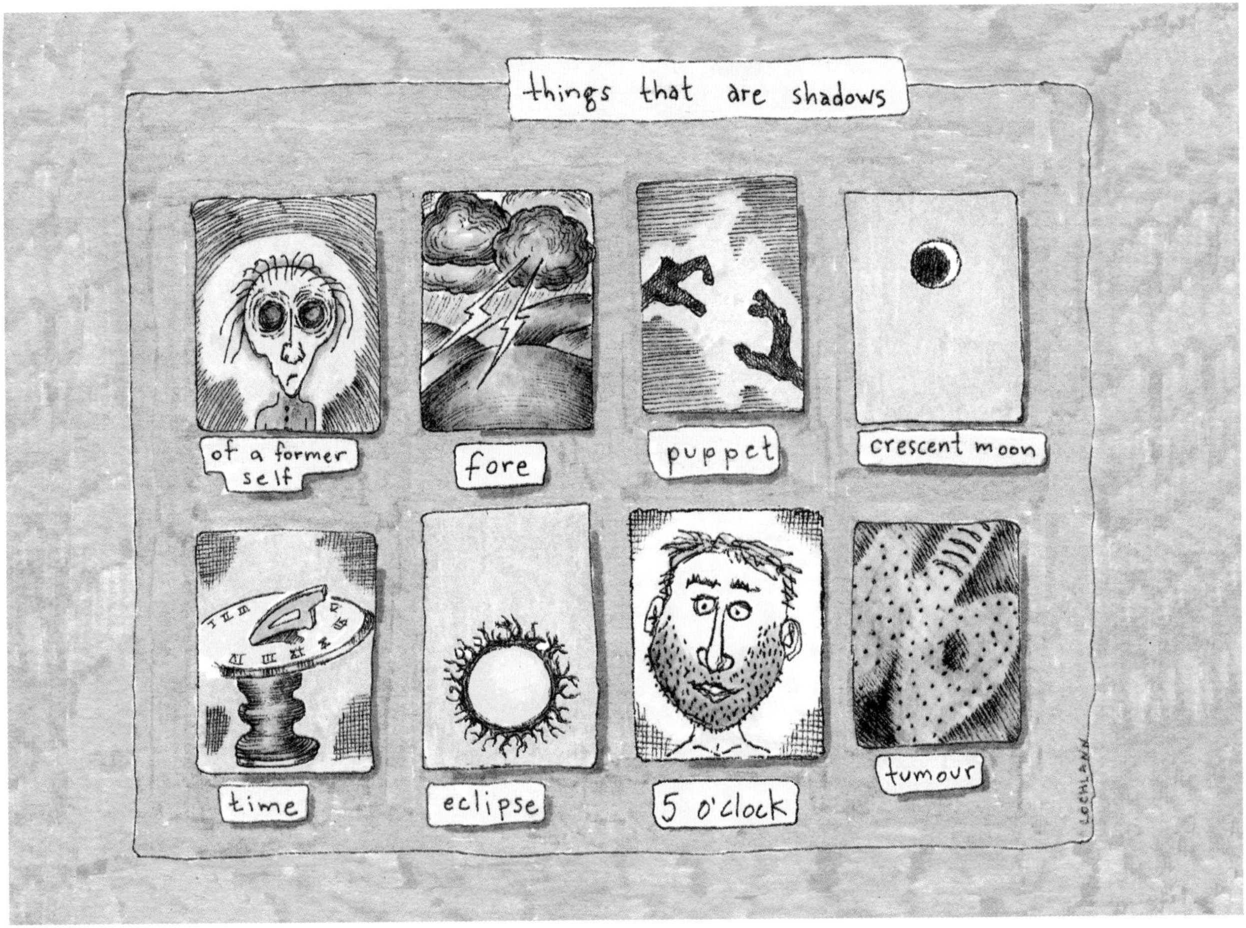
things that are shadows
of a former self
fore
puppet
crescent moon
time
eclipse
5 o'clock
tumour
LOCHLANN

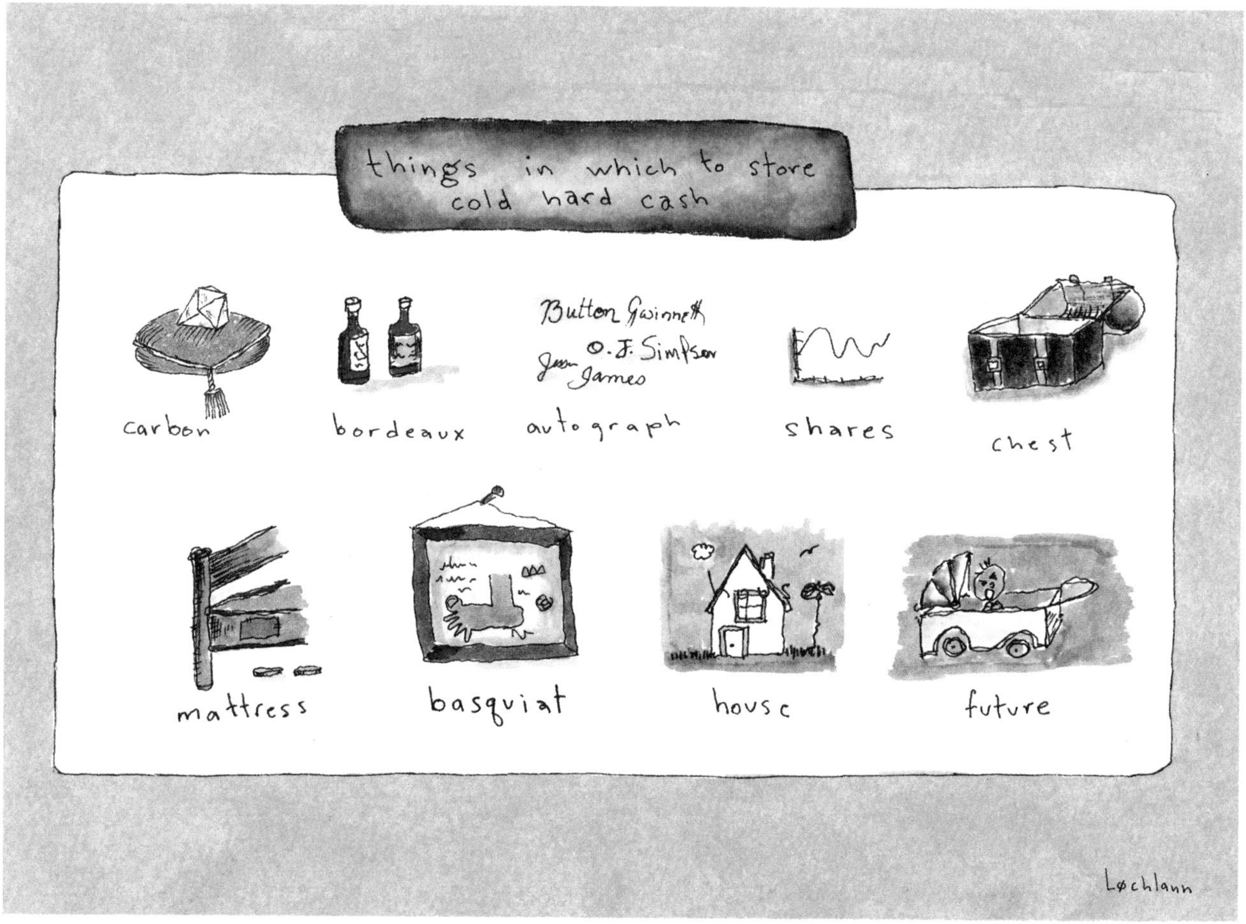
things in which to store cold hard cash
Button Gwinnett
O. J. Simpson
Jesse James
carbon
bordeaux
autograph
shares
chest
mattress
basquiat
house
future
Løchlann

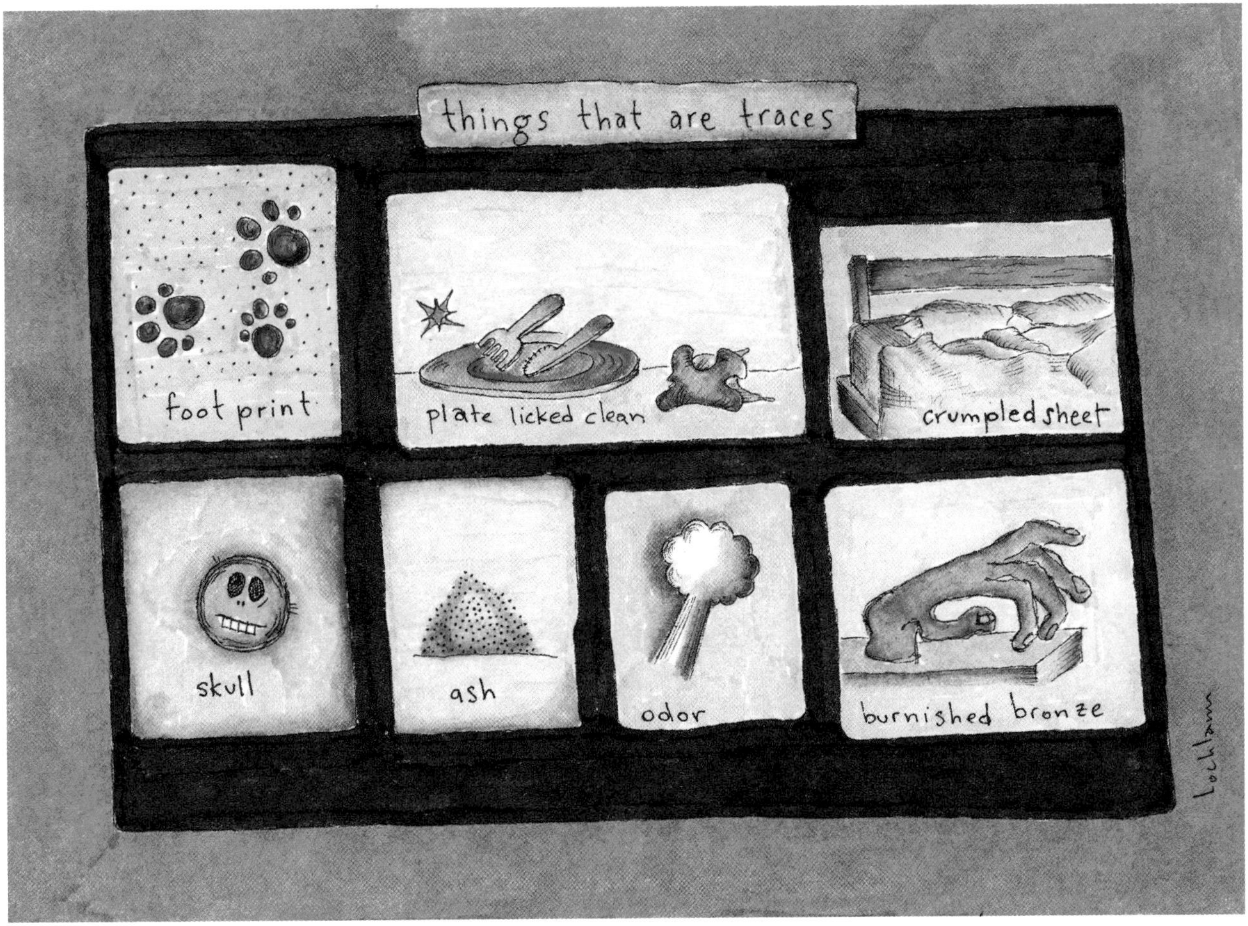
things that are traces
foot print
plate licked clean
crumpled sheet
skull
ash
odor
burnished bronze
Lochlann

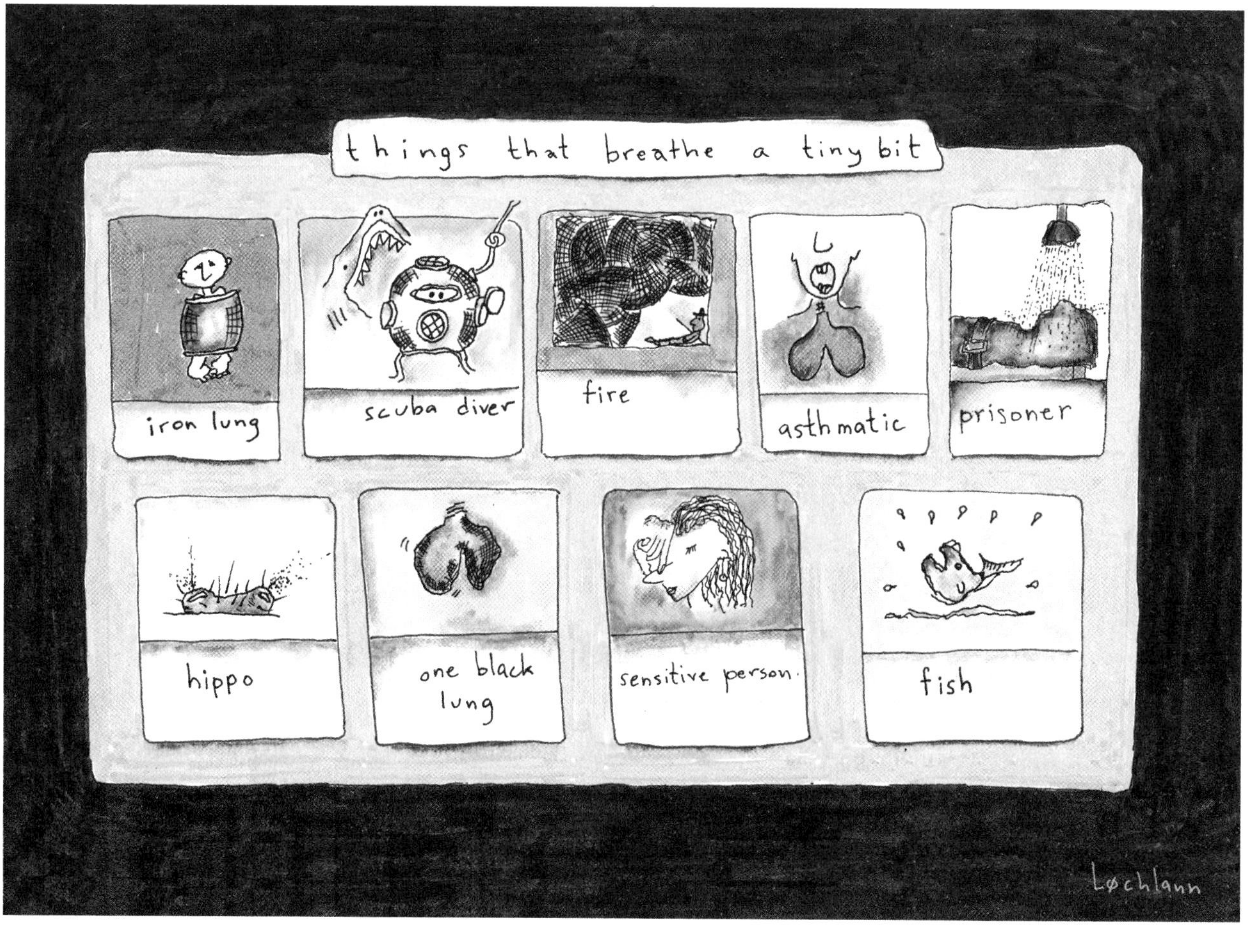
things that breathe a tiny bit
iron lung
scuba diver
fire
asthmatic
prisoner
hippo
one black lung
sensitive person.
fish
Løchlann

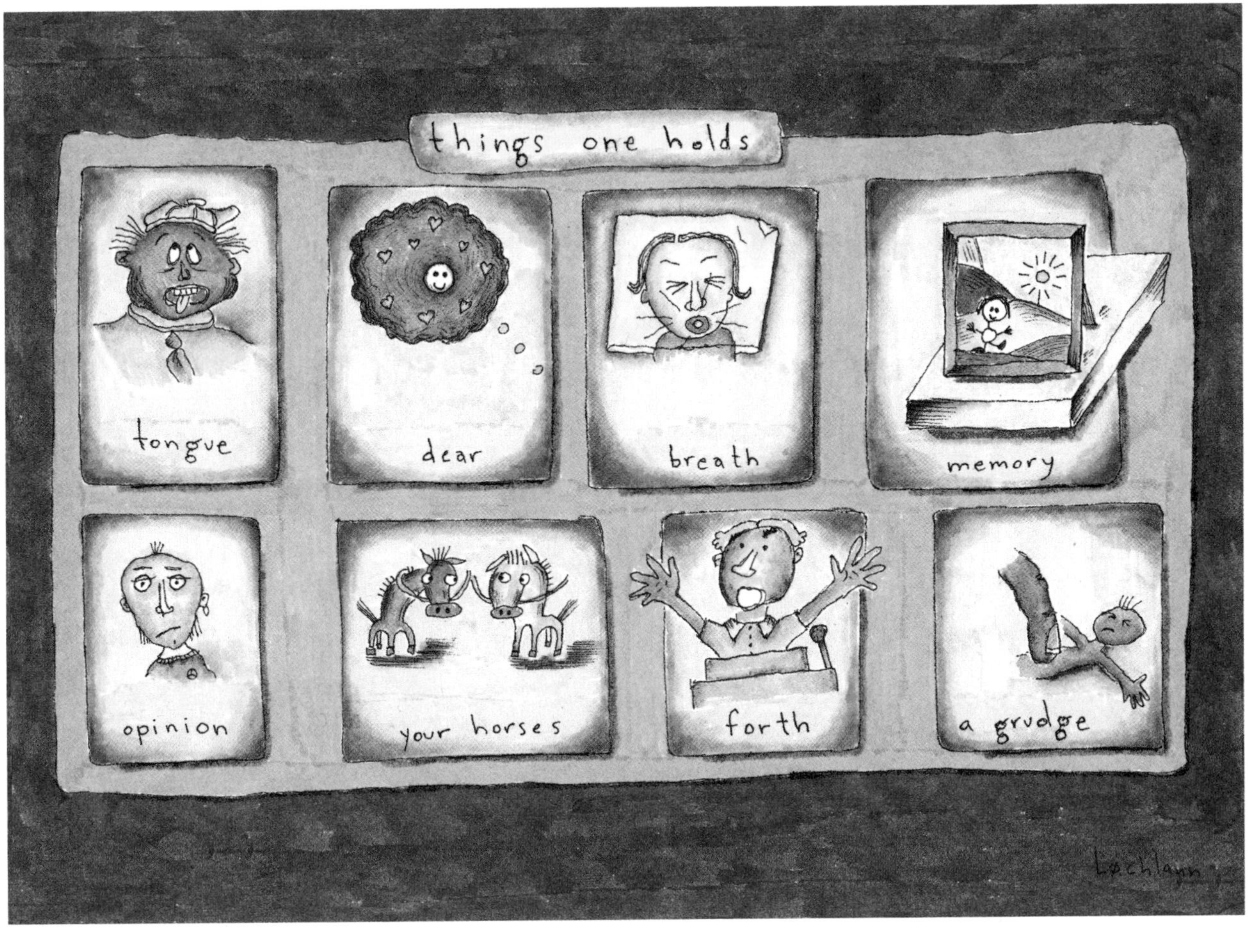

things one holds
tongue
dear
breath
memory
opinion
your horses
forth
a grudge
Lochlann

things that are easily broken, slow to repair

trust

bikini atoll

low blow

1937 Châteauneuf

humerus

crashtestchimp

brain

humpty

Løchlann

things that could describe this onion

skin tone

juice, juicier

body mass index

adhesion

tears elicited

age

feet under

profile

Lochlann

Various Things

MARIA DOLORES McVARISH

Lochlann Jain's artwork addresses a profound conundrum undergirding all conceptual thought.[1] This conundrum has to do with authority, classification, systems of knowledge, and epistemology, among other things. Of its many and varied facets, the one I would like to focus on in connection with Jain's "things that art" series has to do with the roles of identity (sameness) and difference in conceptual thinking. Specifically, I would like to elucidate some of the aesthetic ways in which Jain raises questions about those roles.

By way of introduction, I will discuss a particular "thing that." It's a tiny drawing of an onion with the word "age" written beneath it. Like many of the word/image pairings in Jain's illustrations, it affects me in a peculiar way. What I don't yet know – and what I'd like to find out – is *how* it comes to affect me in the way it does: What is its aesthetic strategy?

One evening not too long ago, I was perusing a selection of "things that art" cards, trying to identify tropes, techniques, patterns, and other qualities that I might wish to explore in essay form. Like others, I noticed Jain's fondness for wordplay ("things that sound a bit like hairspray"), recurring motifs (tongues, knobby hands, ghoulish faces), figures of speech (like "meet your maker"), literalized metaphors ("things for whom the bell tolls"), metonymic displacements (lipstick for lips), and so on. Mildly prudish, I also noticed with amusement the artist's insistence on including things that are normally considered too gross or private to talk about in polite company.

I was perusing, as I say, in this analytic mood when I came across a card entitled "things that could describe this onion." Each of its eight text/image pairings was rendered in oniony colors and looked like it took a little more water than it needed, as a fresh onion often does. Virtually unconsciously, obeying a lifetime of inculcation in Western reading conventions, my eye started in the top left corner and began its left → right, top → bottom sweep. Jain had facilitated this inclination on my part; the top left drawing under the card's title brought me into the series through an eye-catching, clockwise-fanning detail entitled "skin tone." Prompted by its outreaching lines as much as by what I now understood to be "skin tones" in the *next* frame, I shifted right, to an image of two tiny islands in the ocean, very close to one another, occupied by enormous half-onions tipping toward each other. The label said: "juice, juicier." ... Um ... what? I relented in my analytic sweep to consider these words. I understood their utility as "things that could describe this onion" ("juice, juicier" = comparative). But what do they mean as a caption for an image of onion halves on desert islands? From here, confounded, my eye dropped directly down. I'd been trying, somewhat unthinkingly, to "read" (apprehend) the series, but that "juice, juicier" drawing threw me off. Now my eye wanted to travel on its own terms and found a kindred color palette in an image below.

And that is how I stumbled on the text/image pairing that affected me in a peculiar way. Beneath "juice, juicier," I found a thumbnail rendering labeled "age." Here, the aforementioned onion – the one that could be described – sat on a large wooden chair facing a wall-mounted mirror. I stopped short, laughing out loud, but also feeling a tender tug. In the drawing, the onion totally ignores a small, onion-sized door connecting its seemingly windowless domicile to somewhere else (presumably, life outside). A calendar-sized ... um, calendar? ... or framed photograph? ... hangs on the same wall as the door and is equally ignored by the onion. The onion, it seems, only has eyes for its reflection in the mirror, for the slow-motion drama of its own aging process. (Wait, do onions have eyes, like potatoes?) By the time I happened on this card, that onion may have been gazing like this for days, months, possibly years. It was a lonely scene, funny and sad at the same time.

It bears mentioning again that my analytic impetus had, by this time, been thwarted by two giant, juicy onion halves marooned somewhere in the South Pacific. Without that steam shovel in front of me, I was more prone, more vulnerable, more *alive* to the eccentricity of the drawing I now beheld. I was looking at a lonely senior onion whose spouse and friends had probably all died. And if death hadn't yet entered my thoughts consciously, Jain bolstered this association in the text/image pairing immediately to the right: a section view of an onion growing deep underground, labeled "feet under" – as in, *six* "feet under."

I should have continued skimming through the cards, but I preferred instead to revel in my tender feelings for this senior onion. Indeed, the image dogged my thoughts all week. "Ha ha!," I kept half-thinking, more than a little confused: "that's so sad!" – meaning, somehow, that the aged onion had managed to elicit in me a sense of both exhilarating hilarity and depression. Then I went back and looked at the drawing immediately to the left of "age" on the onion card, labeled "tears elicited." It shows, in the foreground, a person weeping over a cutting board with sliced onions. One of their hands, awash in tears, holds the chopping knife; the other raises a handkerchief to staunch the torrent. Jain is throwing me a rope here; I don't have to do all the work to reconcile my feelings about the senior onion. Thanks to the inclusion of "tears elicited" within the "things that could describe this onion," I'm reminded that onions make people cry all the time. Everybody knows that, so by extension, I'm part of "everybody," which can be reassuring when you're wondering about your fondness for a dying onion. On the other hand, onions don't often make everybody laugh. But here, too, Jain has thoughtfully stepped in, offering a way for me to stretch my understanding of what "onion" *does* or *means*, beyond its status as a vegetable, to include the contexts in which we normally engage with onions: in the same tiny thumbnail drawing of the weeping onion chopper, bounding toward the old onion-lost-in-memories next door, is a very jovial person sporting goggles. This character jumps – ta-da! – into (and almost beyond) the space behind the weeping chopper, grinning at but unseen by them.

Let us review: on the left, we have the grin/weeper; in the middle, the emphatically poignant granny onion; and on the right, the onion (six) "feet under." It took me several viewings and a fair amount of writing to make the connections between these drawings conscious. Doubtless, there's much more to appreciate in this card, but I think I have enough to begin addressing the question of how "things that art" work on me, aesthetically, and what this has to do with conceptual thinking.

My first assertion in this essay was that "Lochlann Jain's artwork addresses a profound conundrum undergirding all conceptual thought." In order to support this, I will begin with a bit of theory related to what I believe "concepts" (i.e., components of conceptual thought) *do*. First and foremost, I would submit, concepts *contain*. They gather and represent things that relate to or resemble one another in specific ways. The concept "things lips do," for example, contains, according to Jain: "purse, smack (talk), stick, lick, service, fat, kiss (lock), pucker (pout), whistle, [and] chap." It is very convenient to be able to use "things lips do" as shorthand if I want to think or speak about that topic without having to continually run through particular examples. At a very basic level, concepts help organize the ways we know the world; they aggregate and distill collections of linked items, then reduce those collections to singularities, thereby making normal mental activities like thinking, planning, and remembering – not to mention communicating – possible. Not only do our minds depend on our ability to conceptualize, but we likewise cannot function in society without concepts.

At the same time, however, concepts *constrain*. Just as they aggregate and collect things, so too do concepts put things "in their place," functioning as master terms that force the things they encompass to become subordinate. Normally, the differences between items that fall under one concept must be overlooked, excluded, or denied, and there are practical reasons for this (as noted above). Everything that distinguishes the individuals included in concepts like *we*, *us*, or *our*, for example, would seem to threaten the terms' utility for communication and thought. Yet as "we" all know, inclusion within a we/us/our group always comes at the price of my (or your) originality. Put another way, every concept is constructed over real differences between its component terms, and these differences are suppressed in the name of that concept's identity (singularity). This, in a very general sense, is the profound conundrum undergirding all conceptual thought.

In "things that art," the very premise of identity – that top-down coherence that alone sustains concepts – loses valence.[2] The irregularities, improprieties, and transgressions within each card work both with and against each other, destabilizing the presuppositions of conventional concepts and countering the impetus that encourages us to take them for granted. Jain pointedly acknowledges, tests, and embraces the differences that inevitably reside within concepts (or categories). Indeed, we can hardly make sense of these collections without recognizing the parts that threaten to undo them. And we can't appreciate the *art* of *things that* unless we are willing to include ourselves in what Gilles Deleuze might call the alterity of ideas in Jain's explorations (meaning the occurrence of differences within the combinations and series that constitute "things that art").

Let me elaborate on this notion of including oneself in the alterity of the idea by way of that age/onion "thing" described above. Age is a "thing that could describe this onion," certainly, and if you wanted to use it, you'd expect to talk about when the onion was harvested, whether it had been refrigerated, its water and sugar levels, whether mold is evident, and so on. In other words, you'd expect to use the onion's age to describe the onion, as the card's title suggests. Instead, however, Jain uses an onion to describe age: the attendant image shows someone or something (an onion) looking in a mirror. Now, this might be how one thinks about *oneself* aging – losing mobility and cognition, withdrawing from the hustle of an ever-changing world, growing inward, reflecting (so to speak) back on old times, et cetera – but it's not usually how one thinks about describing one of the characteristics of an *onion*. So, in order to fully appreciate and understand the connection between the concept ("things that could describe this onion") and one of its "things" (age), the viewer must identify with the onion, to some extent, thereby including him/her/themself "in the alterity of the Idea."[3]

This kind of conceptual chiasmus lies at the beating heart of "things that art." A chiasmus is the crisscrossing of a symmetrical verbal structure in which two parallel linguistic forms overlay and then invert one another. In a *conceptual* chiasmus, ideas or concepts normally configured in parallel (the relation of one concept to another, the relation of one subordinate term to another) are similarly overlaid and inverted, thereby collapsing any hierarchical distinctions between them. In Jain's work, this occurs between the supposedly subordinate "thing" and its ostensible master concept at the concept-to-concept level (in the image of the dying onion, which is the master concept, onion or age?). This, I feel, is one way that Jain's art begins to redress the (in)difference of conceptual thought – and as it does, it has a peculiar aesthetic effect (on me, at least).

Another way that Jain responds to the profound conundrum of conceptual thought is through radical, experimental inclusion. In "things that are weapons," for example, the concept *weapon* represents a range of things that have in common a certain instrumentality for hurting or killing. AK-47s, cannons, and poison darts have each been conceived as weapons at different times, in various circumstances. But if we include something in the *weapon* concept that's normally intended for another use – say, cars, for driving – a lot can happen. First and foremost, automobiles are instantly reduced to their capacity to hurt or kill. But the inclusion of cars ripples with more subtle effects and implications as well. Who is hurting or killing whom with this car/weapon? Who – in the chain of car stakeholders, which includes manufacturers, traffic engineers, regulators, and drivers, among others – is responsible for the harm cars bring to the environment, to pedestrians, or to other drivers? This is a topic – a set of questions – that Jain has thought about before.[4] The inclusion of the thing *car* within the concept *weapon* changes how we think about cars and forces us to extend "thingness" to its contexts and uses as well; if a car *is* a weapon, it must meet all the criteria for/as weaponry. By the same token, if the concept *weapon* is to include cars, its criteria will have to be extended to things that aren't necessarily intended for harming or killing. And if that's the case, then the clarity and utility of the concept itself grows weaker. Can *any* thing be weaponized? Is *intent to harm or kill* necessarily a criterion for weapons?

Our practical need for the integrity and clarity of concepts – and particularly for the integrity and clarity of those concepts-in-common that language and culture consist of – forces us to make choices: Is that new or deviant "thing" that threatens the identity of the concept *in* or *out*? Can the concept be stretched or adjusted to accommodate its difference/divergence from a fuller range of related "things"? If so, is the concept bolstered or impaired by that accommodation? And not just in one's personal opinion, but in practice, as currency for exchange with others?

When concepts outright fail, it is because they show themselves to be incapable of managing the differences exhibited by the full range of their supposedly subordinate terms, *except* by means of excluding, denying, or opposing them. Embracing what the concept typically denies (and/or altering the hierarchy between a concept and its constituent things) can radicalize power relationships that extend from all the mental and social structures involved in common concepts. This, I believe, is an esthetic effect of Jain's radical, experimental inclusions.

Because Jain's subject matter is the stuff of culture (language, shared ideas and experiences) both of these techniques – conceptual chiasmus and radical, experimental inclusivity – target subjectivity and, to varying degrees in consequence, the viewer's feelings. "Things that art" can be funny, sober, disturbing, sad, frustrating, and gleeful – sometimes all at once. Indeed, the game is not so much to identify each word/image pairing's outlier(s) as it is to undertake the conscious, often partly *verbal* work of appreciating the ways that each coupling poses its own problems for the titular concept. And that's a beautiful, delightful, and intriguing "thing."

Notes

1 For the purposes of this essay, I use the word "concept" as an analog for "category," hoping to obviate the need for too much vocabulary.

2 **identity**: *n.* one-ness, unity, cohesion, sameness.

3 "[T]he difference is internal to the Idea; it unfolds as pure movement, creative of a dynamic space and time which correspond to the Idea. The first repetition is repetition of the Same, explained by the identity of the concept or representation; the second includes difference, and includes itself in the alterity of the Idea, in the heterogeneity of an 'a-presentation'. One is negative, occurring by default in the concept; the other affirmative, occurring by excess in the Idea." Gilles Deleuze, *Difference & Repetition*, trans. Paul Patton (New York: Columbia University Press, 1994), 24.

4 Lochlann Jain, "'Dangerous Instrumentality': The Bystander as Subject in Automobility," *Cultural Anthropology* 19, no. 1 (2004): 61–94.

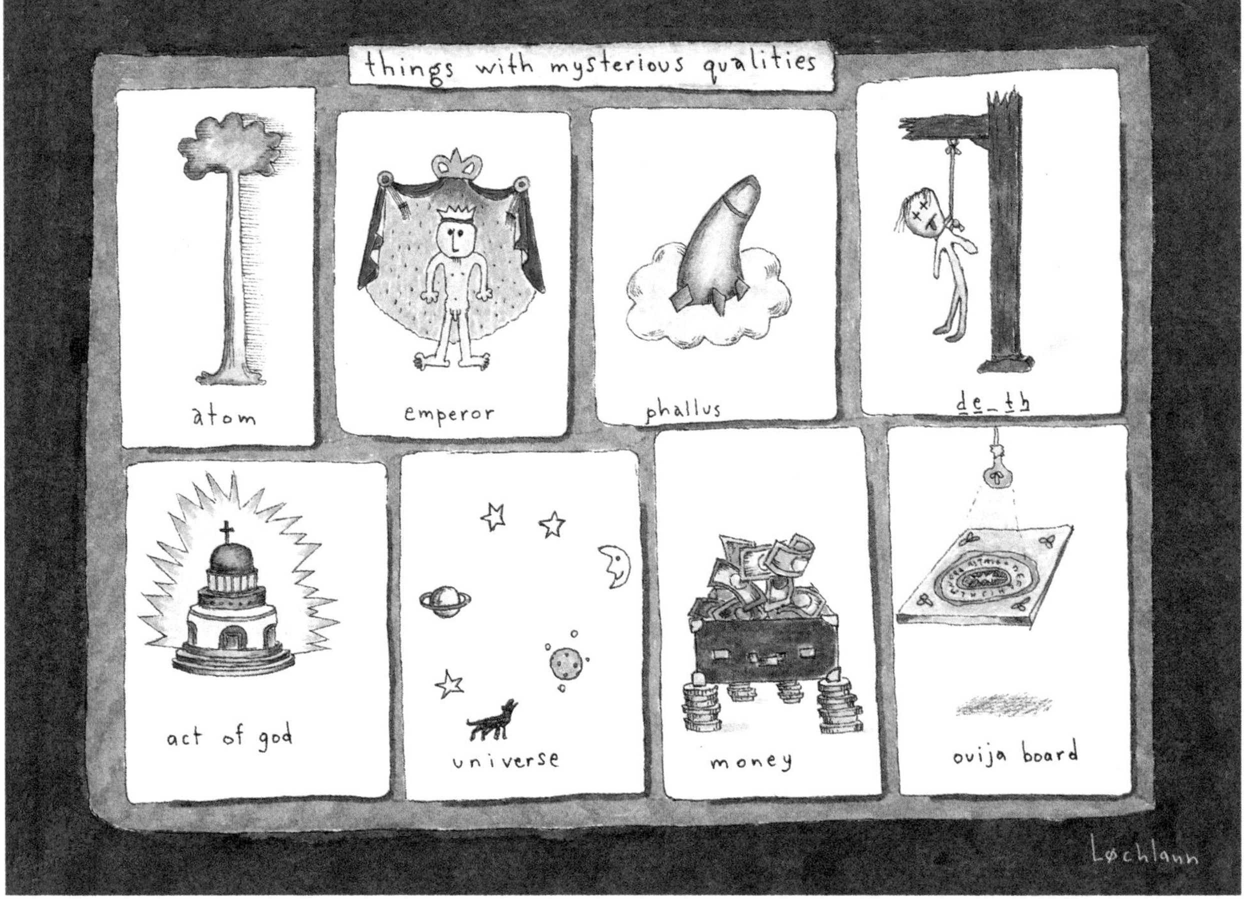
things with mysterious qualities
atom
emperor
phallus
de_th
act of god
universe
money
ouija board
Løchlann

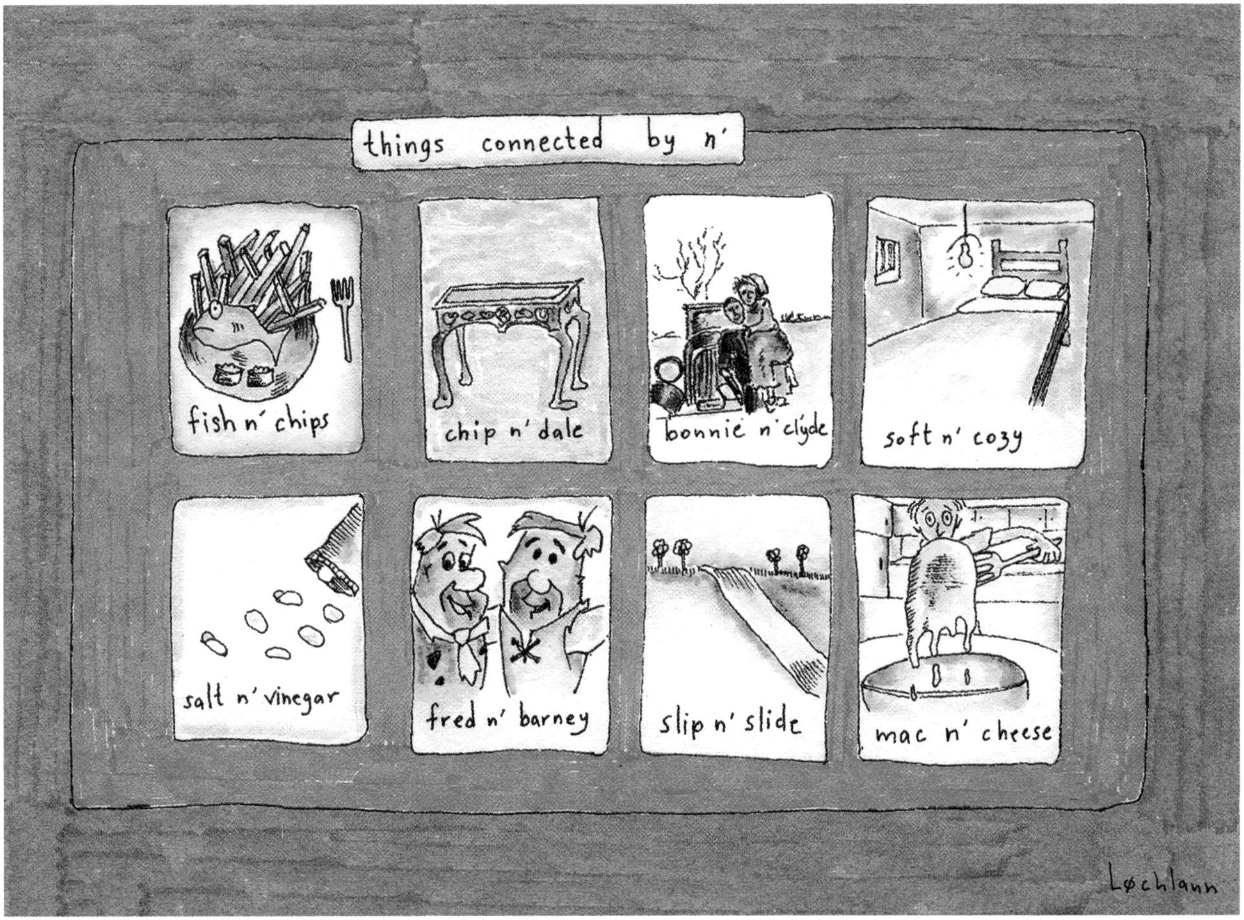

things connected by n'
fish n' chips
chip n' dale
bonnie n' clyde
soft n' cozy
salt n' vinegar
fred n' barney
slip n' slide
mac n' cheese
Løchlann

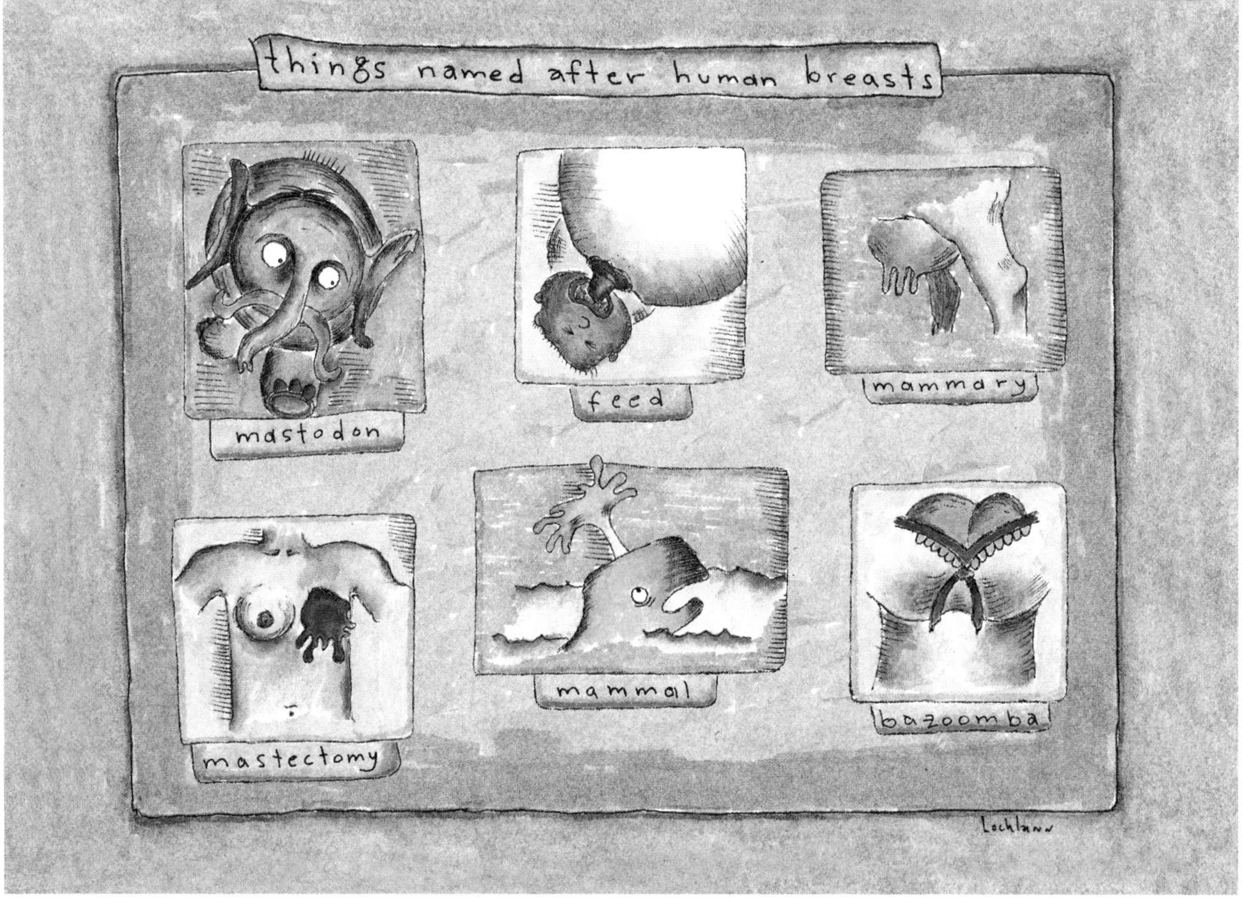
things named after human breasts
mastodon
feed
mammary
mastectomy
mammal
bazoomba
Lochlann

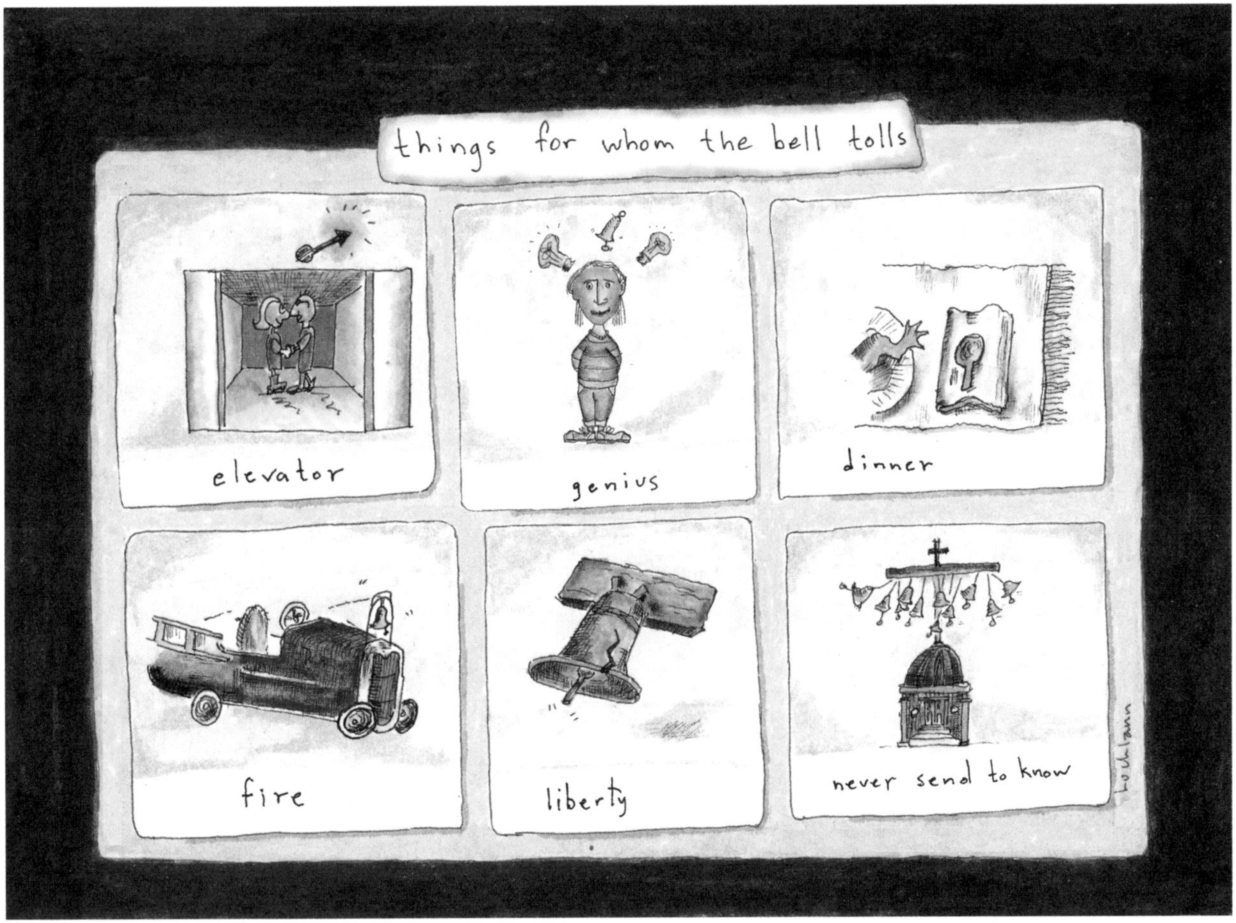
things for whom the bell tolls
elevator
genius
dinner
fire
liberty
never send to know
Lochlann

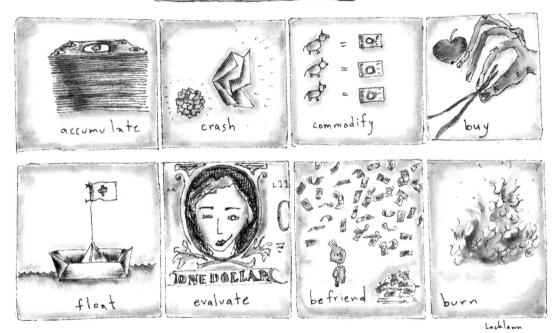

things a dollar does

accumulate   crash   commodify   buy

float   evaluate   befriend   burn

Lachlann

things that will happen when

hell freezes over

father comes home

meek inherit the earth

workers lose their chains

the fat lady sings

your time has come

Løchlann

things recommended not so long ago for the resuscitation of the drowned

induce vomit

smoke in anus

bloodlet

bellows

sheep skin cover

hang

Lochlann

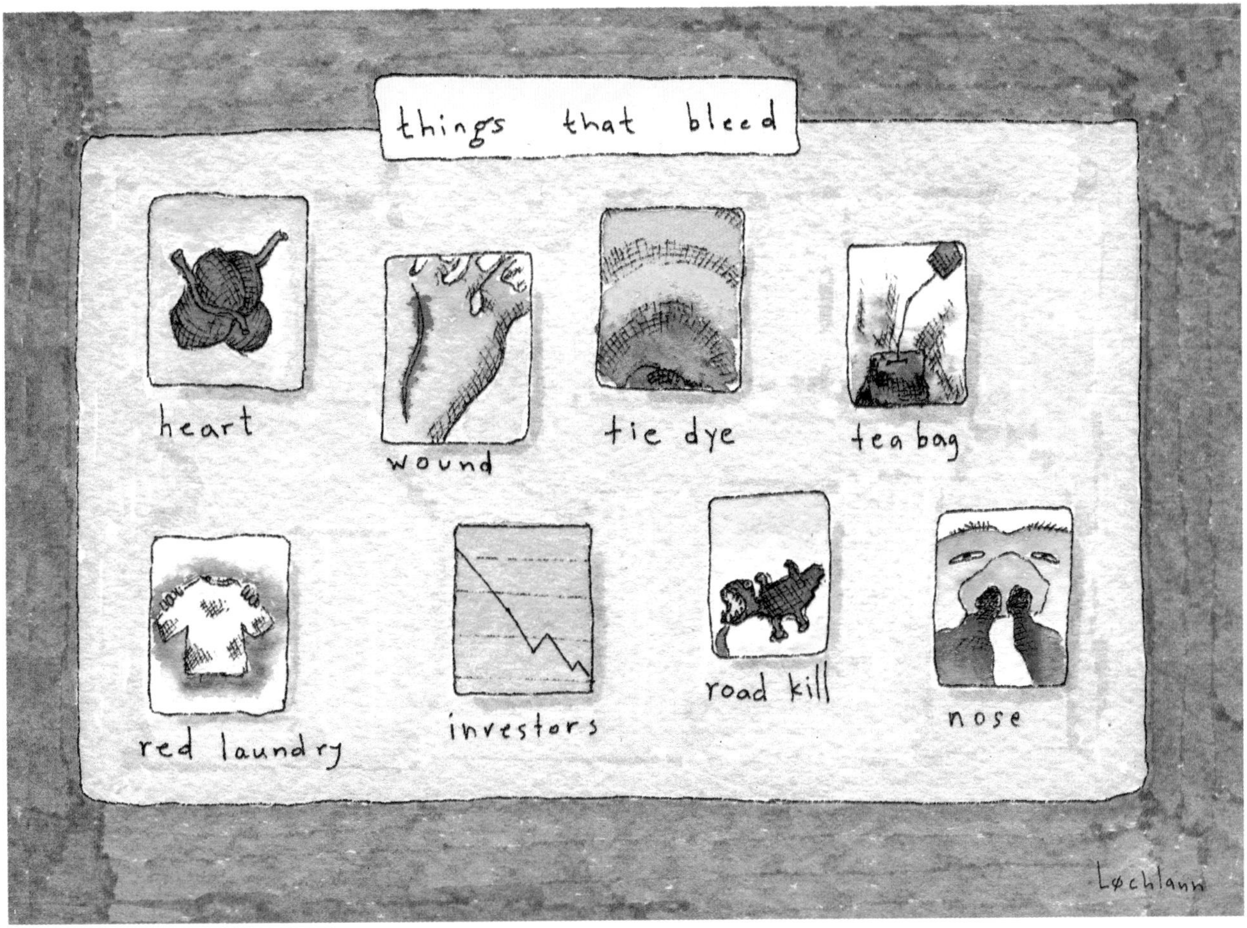
things that bleed
heart
wound
tie dye
tea bag
red laundry
investors
road kill
nose
Løchlann

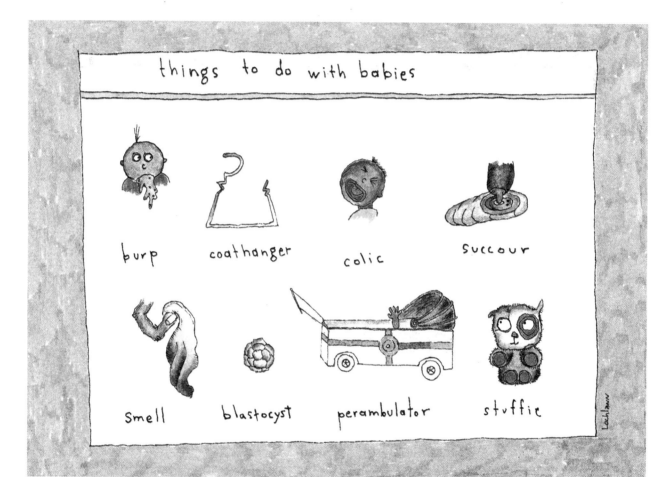

things to do with babies

burp

coathanger

colic

succour

smell

blastocyst

perambulator

stuffie

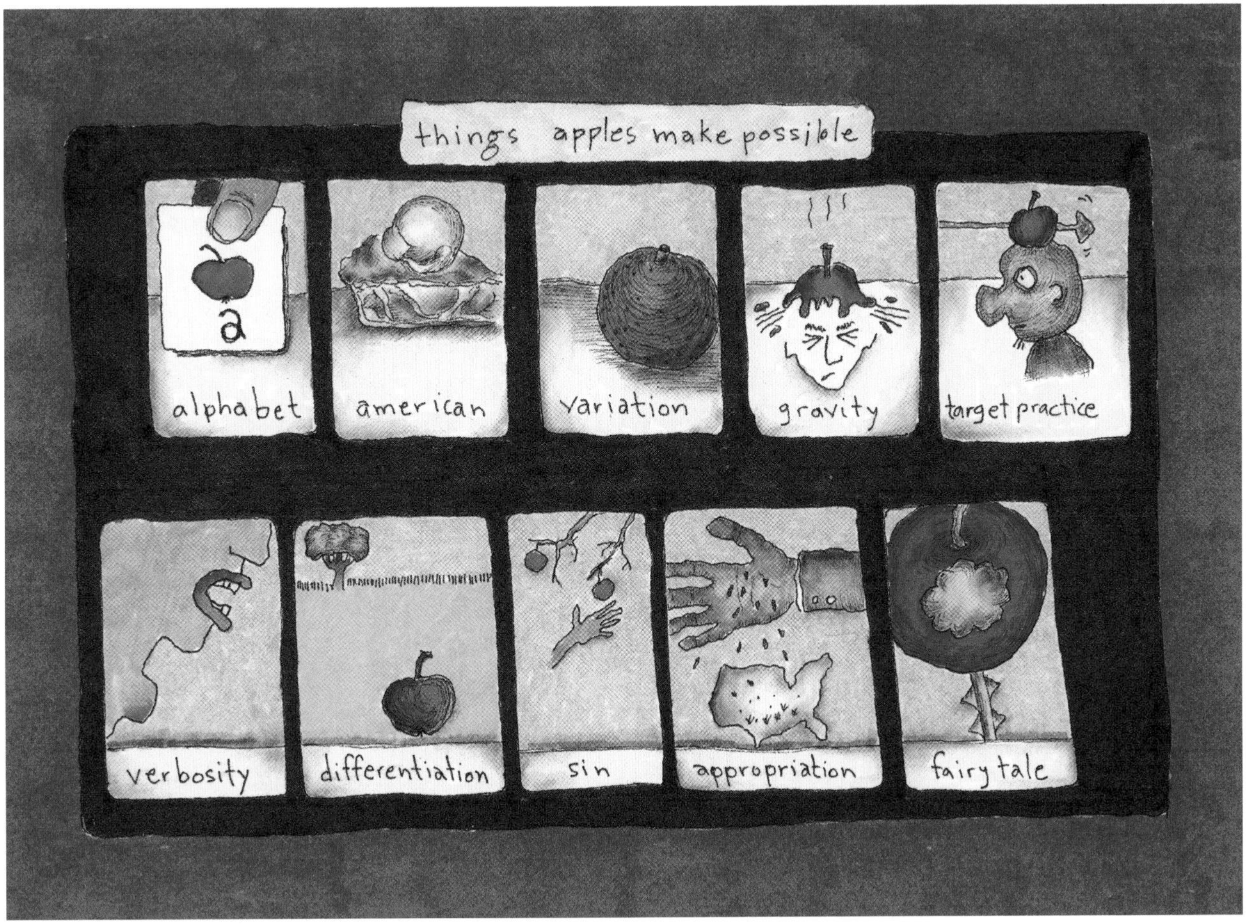
things apples make possible
alphabet
a
american
variation
gravity
target practice
verbosity
differentiation
sin
appropriation
fairy tale

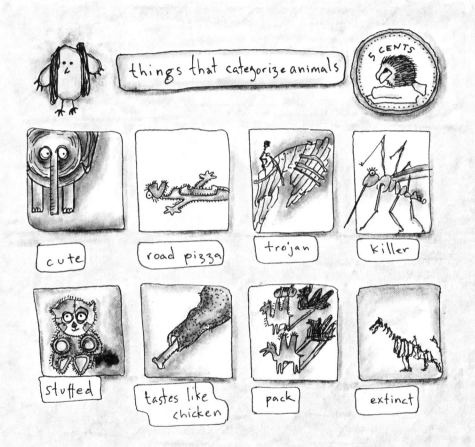

things that categorize animals

cute

road pizza

trojan

killer

stuffed

tastes like chicken

pack

extinct

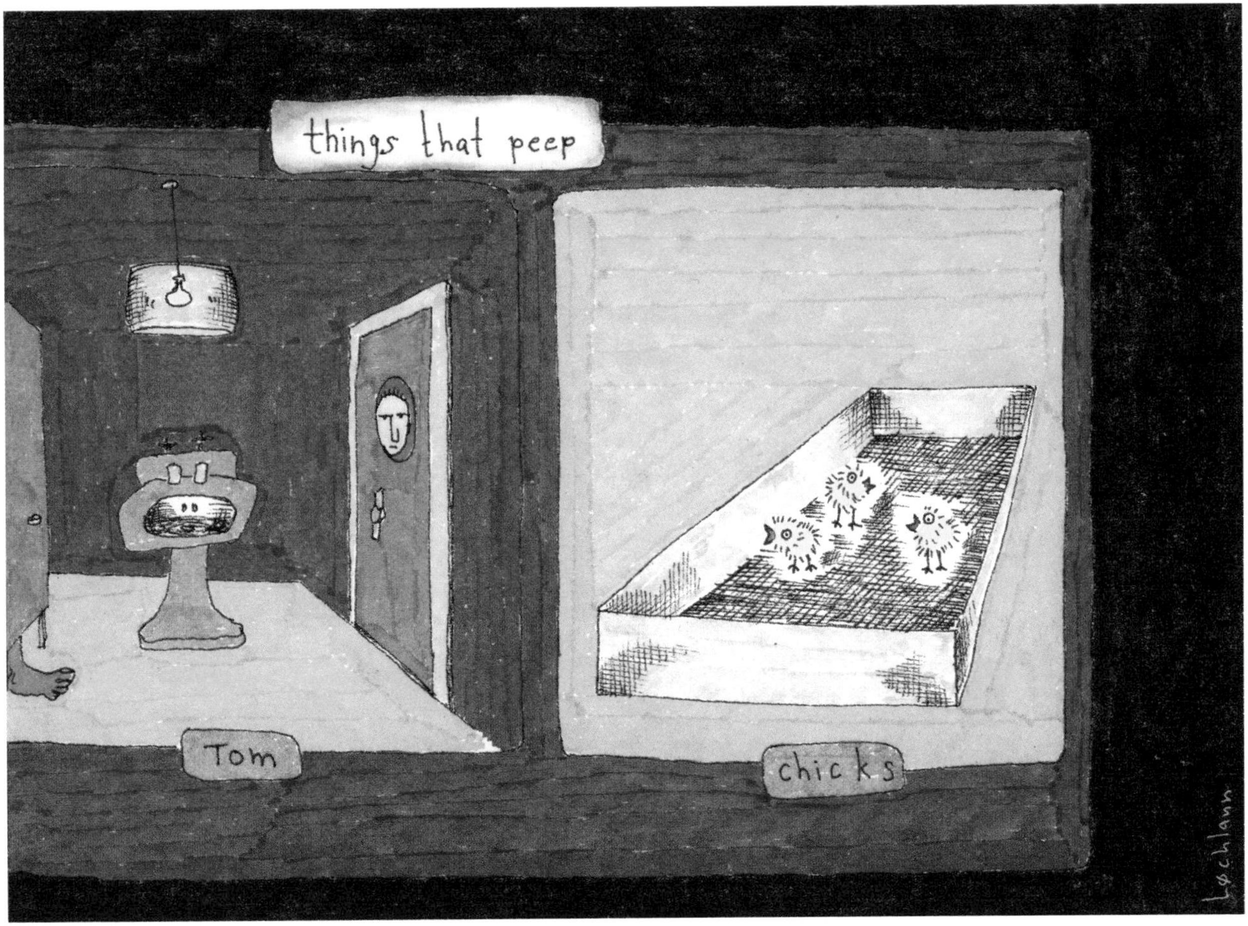
things that peep
Tom
chicks
Lochlann

"Natural" Collections:
The Whole, the Sum of the Parts

ELIZABETH BRADFIELD

The rustle of flashcards created for study; the tidy drawers of a now-old-fashioned library card catalog; small contextual notes mounted alongside paintings in museums; even postcards, those attenuated missives that stand in for miles and miles of travel and experiences ... all serve as simplified maps of understanding and records of contemplation. The physical size of these objects is part of their power, large concepts reduced to the size of the palm. All are in resonance with Lochlann Jain's cards in *Things That Art*, but none more, to my mind, than the specimen cards in an old-fashioned natural history museum, preserved in its full wondrous and disturbing Victorian aesthetic.

Take, for example, the Wagner Free Institute of Science in Philadelphia: three stories of specimens – birds, mammals, bivalves, minerals, fossils – stored in wood-framed glass cases. Dark oak-paneled walls, an open center of tall cabinets ringed by balconies with yet more specimens, everything gathered and yet separated. Cataloged.

You bend to the crustaceans cabineted at waist level, glass slightly canted to avoid glare (yet there still is glare). You stoop to the faded tests of crabs, the yellowed labels marking species, collector, place, and date of collection. Some are written in slanted calligraphic script, some typed (imagine the satisfying metallic clack of keys, the tick-tick as paper is rolled up line by line). There are drawings, too, to help parse the more esoteric or tiny: a walking-stick bug's coxa and thorax, the strange architectural wonder that is the apparatus of an urchin's mouth. Parts isolated from the whole, offered for singular contemplation. And in cabinets with multiple examples of the same specimen? A testament to variation, to the dubious nature of a singular ideal form.

Peering closer, the biologist or naturalist might note, too, errors of classification on the labels. This gastropod is no longer in that genus. Now this bird is in another family altogether. There is authority in the hand that grouped these creatures originally, yet time has unveiled different relationships, has shaken certain assertions and revealed new, different, or hidden connections. And from those new understandings come also practical apprehensions of how grouping or splitting creatures influences legal decisions about hunting, management, and habitat.

Consider that in the late 1800s, C. Hart Merriam, the "father of mammalogy," divided brown and grizzly bears into 86 separate species where now science names just one – *Ursus horribilis*. How might you offer hunting permits for 86 species or "manage" them? What does "extinct" mean, both to us emotionally and also to the world's biodiversity, when the last of the 86 species is lost? Is there less impact when it is not a species, after all, but an individual that is gone? Why? The names matter. The way we see things in relation to others matters.

Lochlann Jain's categories in *Things That Art* invite contemplation of just such histories: looking, collecting, and labeling. Jain's paper cabinets summon us with sly humor and unconventional logic to question typical groupings and categorizations. Take, for example, "things that are shadows," which includes the scruffy beard growth of "5 o'clock," "time" as represented by a sundial, and "tumor" as an x-ray's blurred finding. The way these "things" leap between different emotional registers unsettles us, and the wit behind the part-phrase "of a former self" as one aspect of this category delights, allowing one to finish the saying and thus feel in on the joke. Together, these examples of "things that are shadows" disrupt a familiar word and invite readers/viewers to ask themselves what is "shadowy" to them. There is an energy, a shimmer. Deliberate gaps between word and image enliven both.

In Jain's work, both word and image vie for a hold on "truth" and we, the viewers, realize that our true interest is in the energy of the varied and discordant claims alongside one another rather than any settled determination. Jain doesn't explain why particular objects are selected for a given category or why the drawing "of a former self" is so tortured and Munch-like while the "crescent moon" is so serene. We are left to work that out ourselves, to determine where the emotional center of the grouping might be found. In this way, Jain's work becomes not illustration/caption, but art. Illustrations bestow simplified clarity. Captions present explanation. Art, however, offers gaps, which the reader/viewer must actively hurdle in order to find and create meaning. It's a thrilling leap.

When we visit a Victorian natural history museum now, it's both *what*'s shown and *how* it's shown that we wonder at. The mind and spirit behind the collection, too, shuffle under our awareness – uncomfortably so. We know this is wrong, this enthusiasm to collect unchecked, these animals removed from habitat and context. How does the beautiful, sterile presentation ignore the people who walked the lands from which the specimens were taken? How does it ignore the mechanism of collectors getting *to* their point of collection? Class, race, and gender are at work here. (Who could afford to travel and collect and display? Who was trusted enough to be believed when they presented the first pineapple as an edible fruit?) Colonialism's legacies are clear, as is the dangerous whiff of exoticism. Jain's work evokes this world and this type of looking, as well as a challenge to it. Each card invites us to consider things as a "type," and also to consider how ridiculous "type" can be, how dangerous. What is a "typical" fruit, emotion, association, human, gender?

Art asks us to think about the nature of truth as well as how slanted and personal knowledge really is. To do this with humor, as Jain does, creates an open invitation to more readily grapple with difficult categories and truisms thrumming in the world beyond the book. Take "things some people jolly well eat but we tend not to," which includes a rather alarmed horse, a surreal knuckle sandwich, crow, and an unappetizing heap of lard. Behind the amusement of "crow" lurks the cultural question of who "we" might be. People *do* eat horse, marrow, lard, and crickets. They also eat crow. They eat dust. Jain has accomplished a subversive sleight of hand: by focusing on what is/isn't eaten, we ultimately come into awareness of cultural identity and the authority to judge what is appropriate or desirable.

Another example of Jain's sly countermelody is the card titled "things that were," which fully delights in illogic in its sparely drawn moments. Is a baseball through a broken window inevitable? No. Yet, to those who have witnessed or caused such an event it could feel that way. And what about the cocaine in the early formulas of Coca-Cola, the beverage that illustrates "hidden in plain sight"? This example prompts us to recall other things once considered benign that we now must question: DDT thinning the eggshells of bird species, microplastics in the ocean, institutionalized racism. Again, it is the quirky and individualistic assemblage in each "things that ..." which invites us to assess what we might add to the category. Rather than building our defensive walls and resisting an explanatory voice telling us "what is," Jain's work enlists us as coconspirators in creative questioning.

The formal elements of *Things That Art* also call to mind the simplified illustrations of cells in biology textbooks. The nucleus and vacuoles, the centrosome and mitochondrion, all those isolated parts working within the cytoplasm. Perhaps it's the tidy bubbles around each example, or perhaps it's because the body is so often evoked in Jain's drawings – the vulnerable, naked penis; the vulnerable, naked nostril inhaling whatever floats by, unable to close anything out. There's a tension between the many neat rounded squares encircling the "things that" which spotlights and isolates them – and perhaps protects them, too, as a cell wall does. These drawings are living things. They are cells going about their work and creating new energies, new processes – organisms in which complexity is enabled by multiplicity. One cell does a single, remarkable task. Together, many cells create infinite and glorious variety.

And, I must ask, is there an element of voyeurism in all this looking? Of peeping in (as through a window or a microscope) on something ongoing not intended to be witnessed? Do Jain's drawings make us culpable, in some sense? Each page contains little portholes, windows onto the mind's inner associations, little ships of an idea afloat on the page's sea. Yes, it's like that. All the elements of each card like the portholes on a ship, people seen in their cabins as if by someone out on the ocean itself, a disembodied watcher. Who is the one watching?

The one watching, of course, is the one whose pen is at work, drawing and shading what is both inside and outside those windows, those cells. One of the powerful elements in *Things That Art* is the time evidenced in the overlapping marks of the pen in those shaded exteriors on so many of the cards, such as "things you may kiss," and "things that abstract."

In looking at Jain's work, I imagine the human bent to the page, repetitively scratching the surface, considering what is revealed in that unearthing. Time and the evidence of its passage moves "things that" from play to study. What happens over the course of all that making? Someone is thinking. Some mysterious linkage emerged over the course of drawing to lead the artist from one cell/example to the next. Would the leap have been as strange or wonderful without the contemplative time it took to create the first example? Someone has pondered and drifted and examined and saved a record of that time – time and associative thought that is unique to the artist's own imaginative and experienced world and thus a gift to us as viewers.

In gathering and preparing specimens for a natural history museum, time is marked by the geographic range and local difficulties of collecting. It takes time to go out and net all those butterflies (think of those that escaped, those that were harmed beyond salvage in their capture; think of the "things that" which were discarded because of the physical parameters of the small card). It takes time to box the specimens and ship them home (think of the steady care it takes to draw a clean, delineating line around each example), and time to mount and group the specimens in some pleasing and logical order (time to plot the final element of the category, to decide when to inject humor, when to prod a bit more pointedly). The speed of our digital worlds, the automated tasks that run behind our daily activities, whether it's typing an e-mail or snapping a photograph on our phone – technology erases the complexity these tasks necessarily embody and emerge from. The evidence of the hand in both Jain's writing and drawing reminds us to consider another pace of contemplation.

Finally, in aggregation, there is power – both within the cards and between them. The largest biological organism in the world is an aspen grove. Each tree is not "a tree," but clonal, an expression of an interconnected whole. Or consider organisms such as salps: small, barrel-shaped, gelatinous marine organisms that have both an individual life phase and a communal one in which they form long chains. Both phases are necessary. Both have different requirements and existences, yet both are "salp." But banded together as grove or chain, we notice them; not just the group, but also the variation between individuals and the amazing cohesion of the whole.

So, too, in Jain's work – to understand the power of what is being rendered, we must hold each drawing on a card and the card as a whole; we must hold each card and also the dance between all the cards themselves, as well as our experience as we turn the pages back and forth, drifting along, following threads of association and inspiration. We need the isolation of the individual, the complexity and strength of aggregation. We need certainty and uncertainty. We need serious play to open a door to a more dangerous consideration of the waters through which we sail.

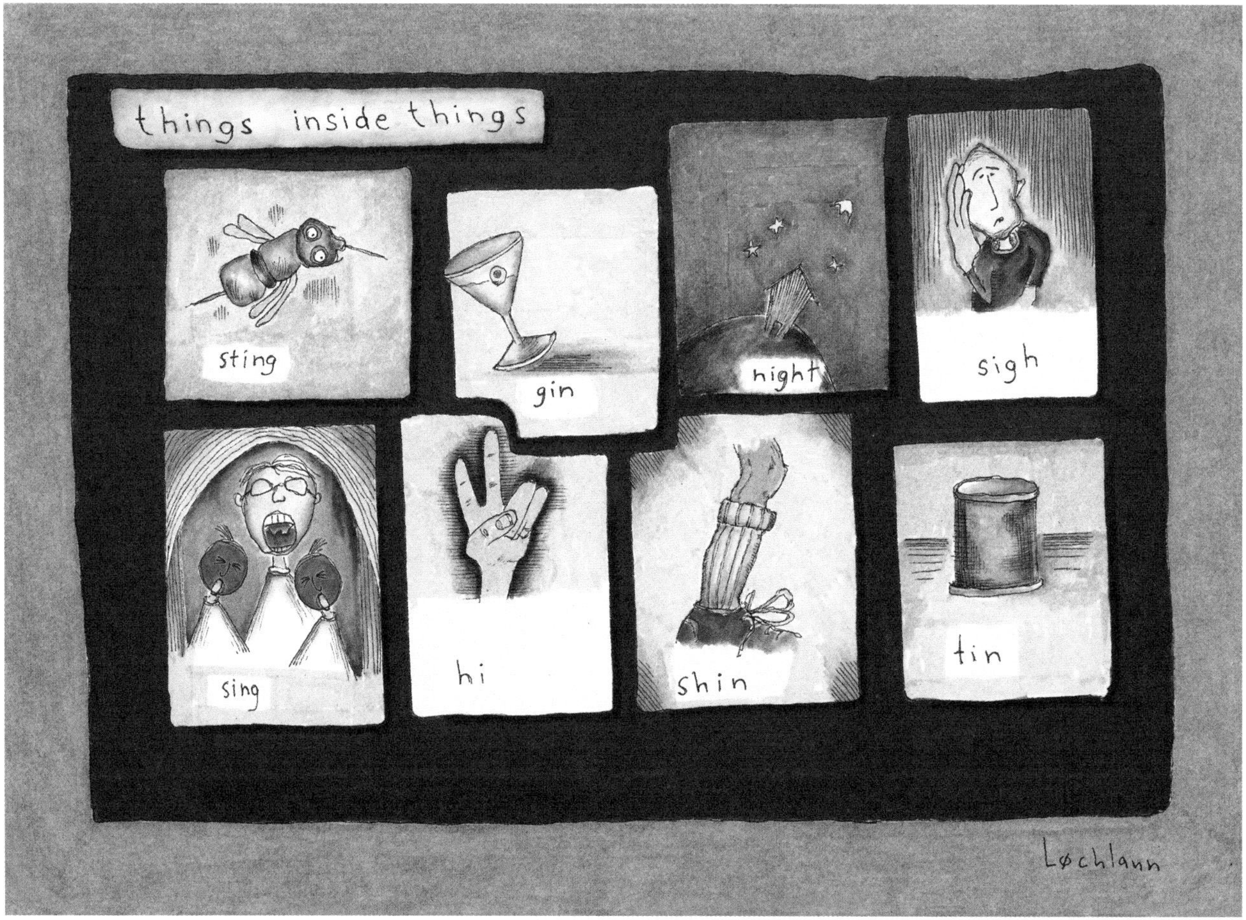
things inside things
sting
gin
night
sigh
sing
hi
shin
tin
Løchlann

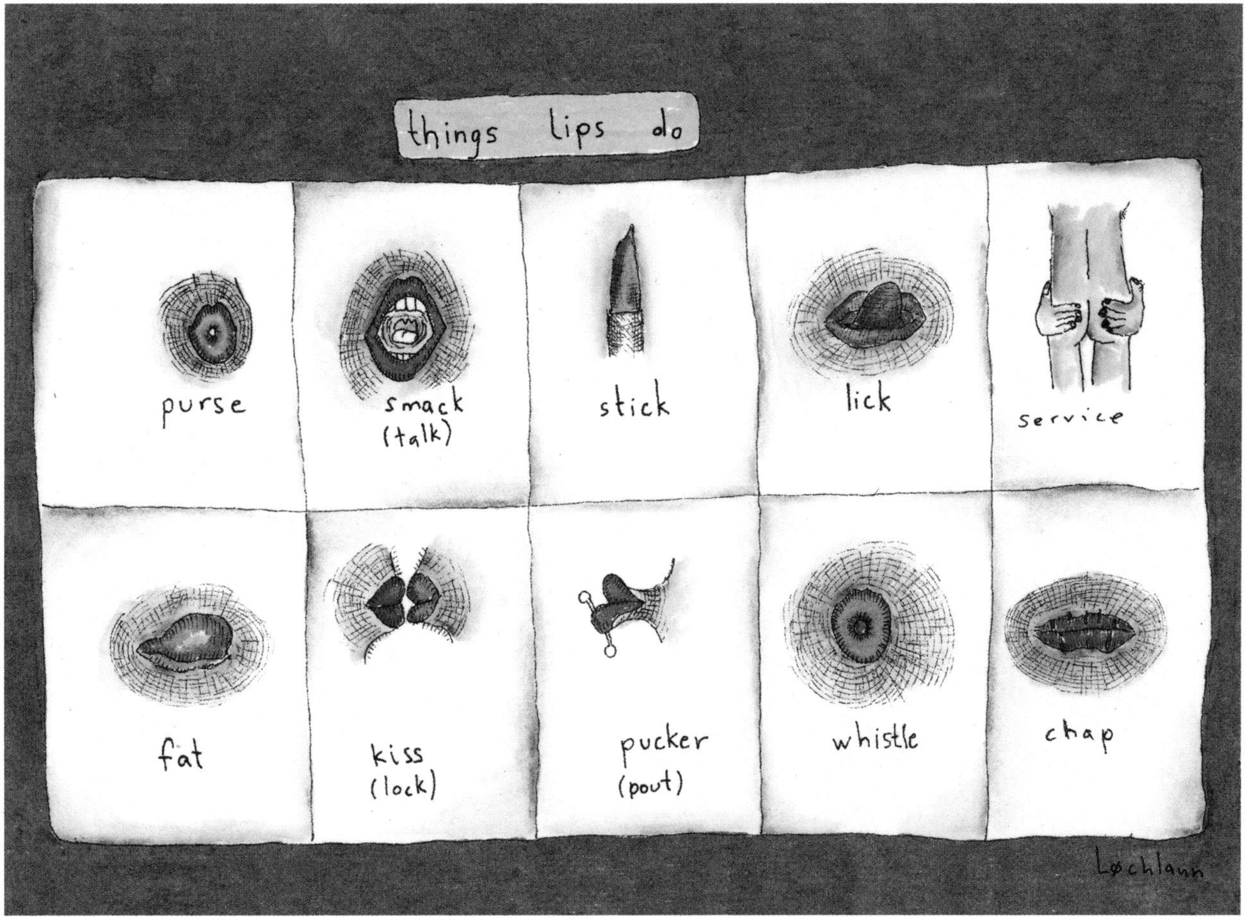

things lips do
purse
smack
(talk)
stick
lick
service
fat
kiss
(lock)
pucker
(pout)
whistle
chap
Løchlann

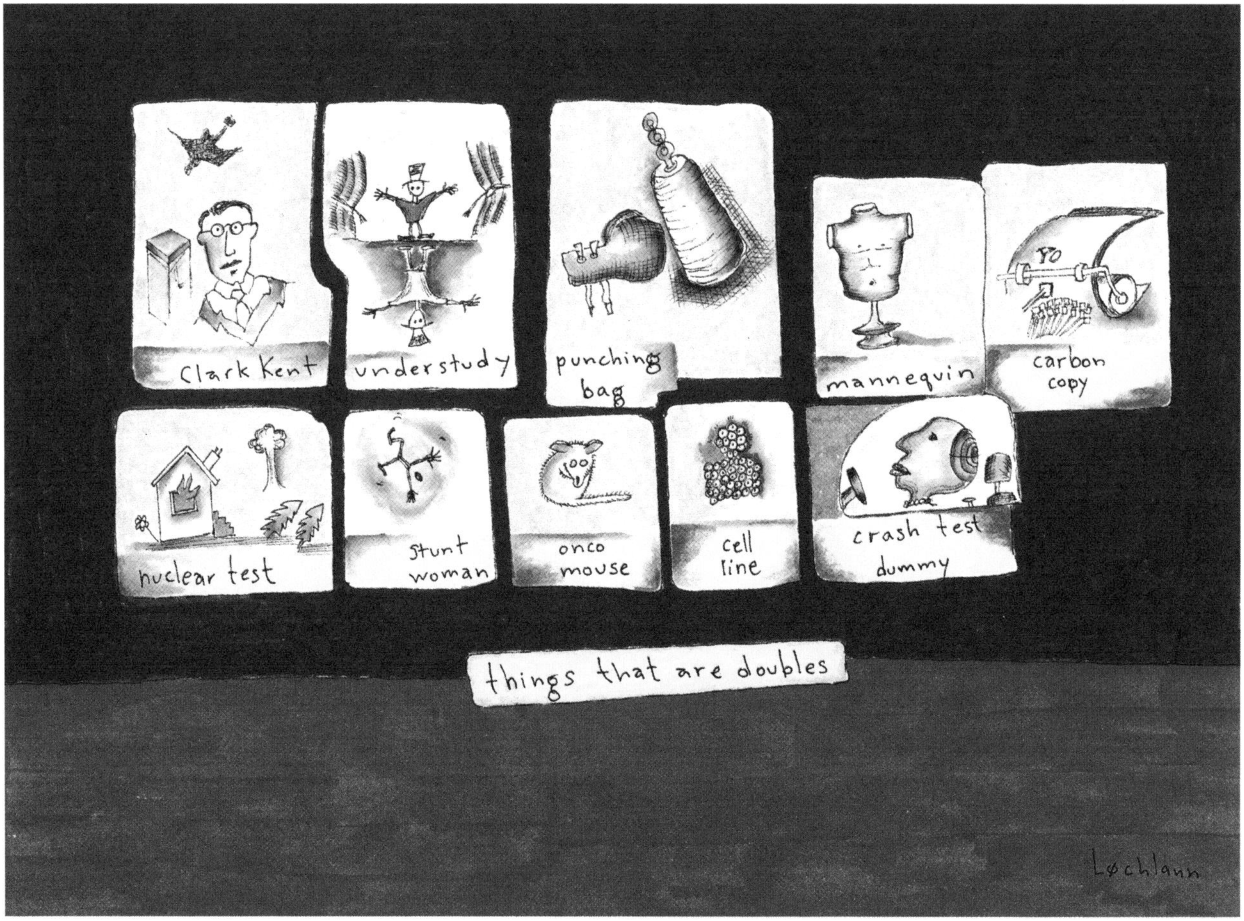
Clark Kent
understudy
punching bag
mannequin
carbon copy
nuclear test
stunt woman
onco mouse
cell line
crash test dummy
things that are doubles
Løchlann

things not valuable in themselves
but as signs of other things

money

clothes

degree

punctuation

manners

gender

ring

good looks

Løchlann

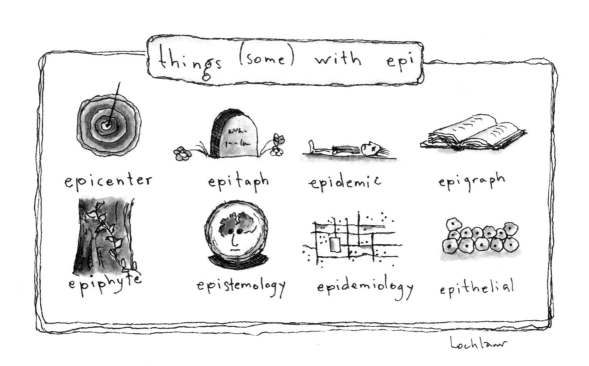

things (some) with epi

epicenter epitaph epidemic epigraph

epiphyte epistemology epidemiology epithelial

Lochlann

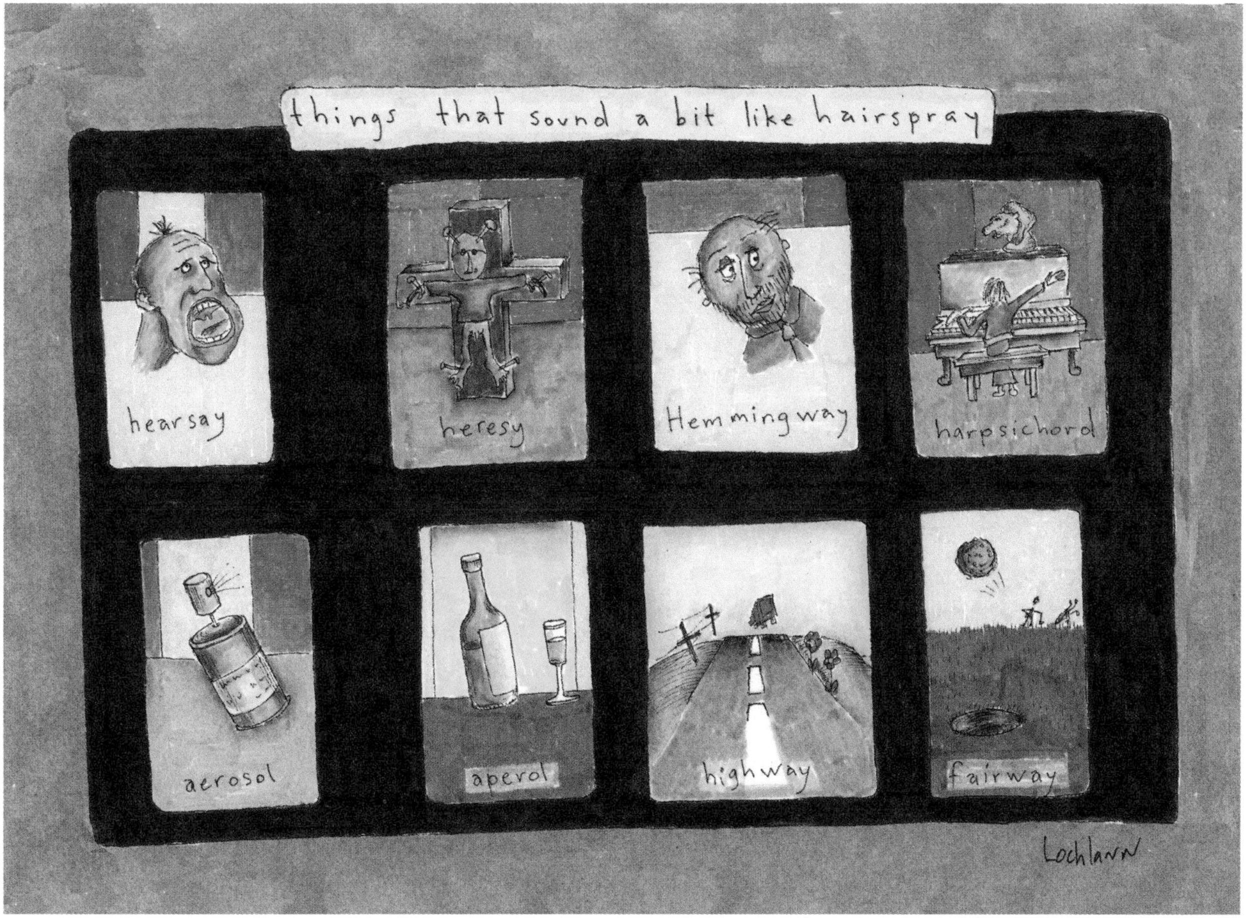
things that sound a bit like hairspray
hearsay
heresy
Hemmingway
harpsichord
aerosol
aperol
highway
fairway
Lochlann

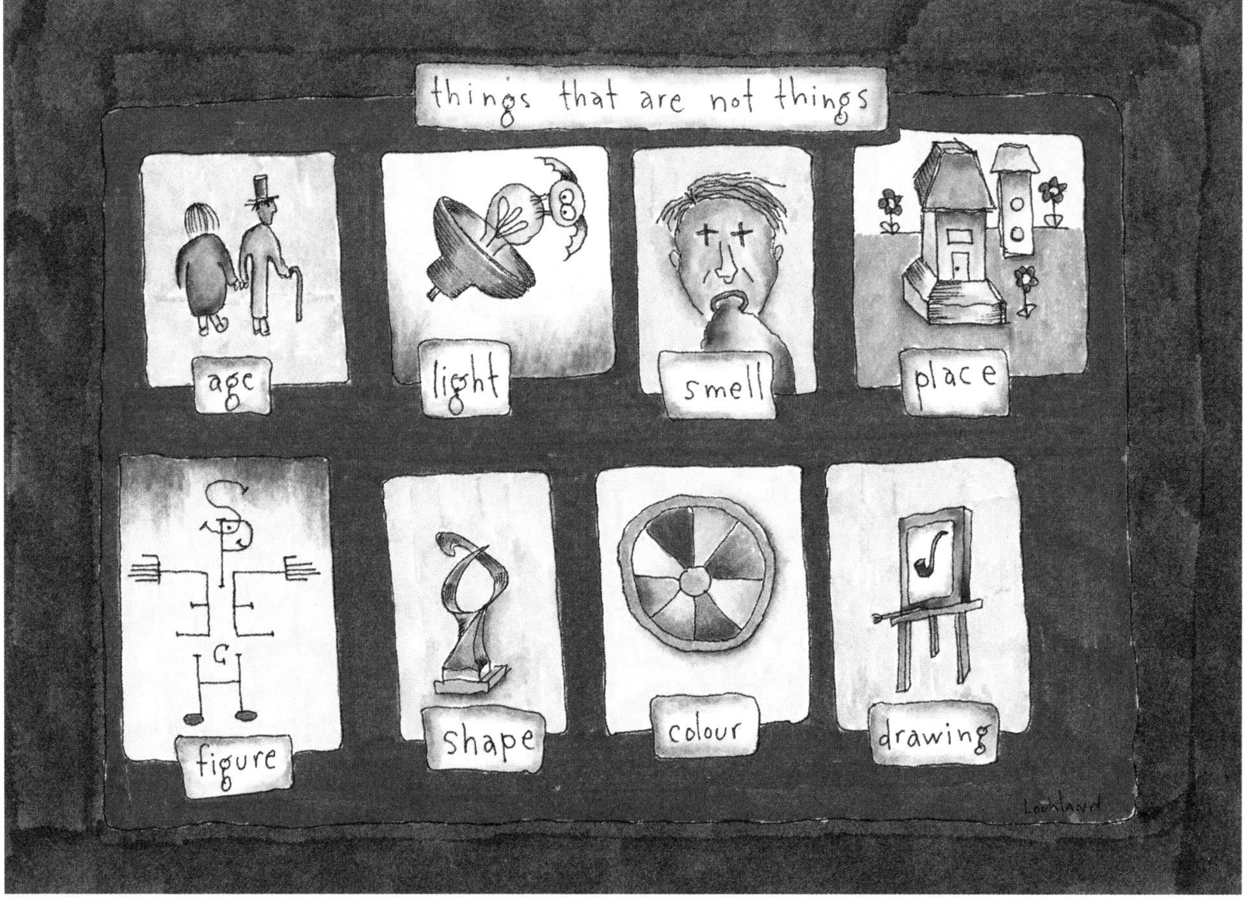
things that are not things
age
light
smell
place
figure
shape
colour
drawing
Lochlann

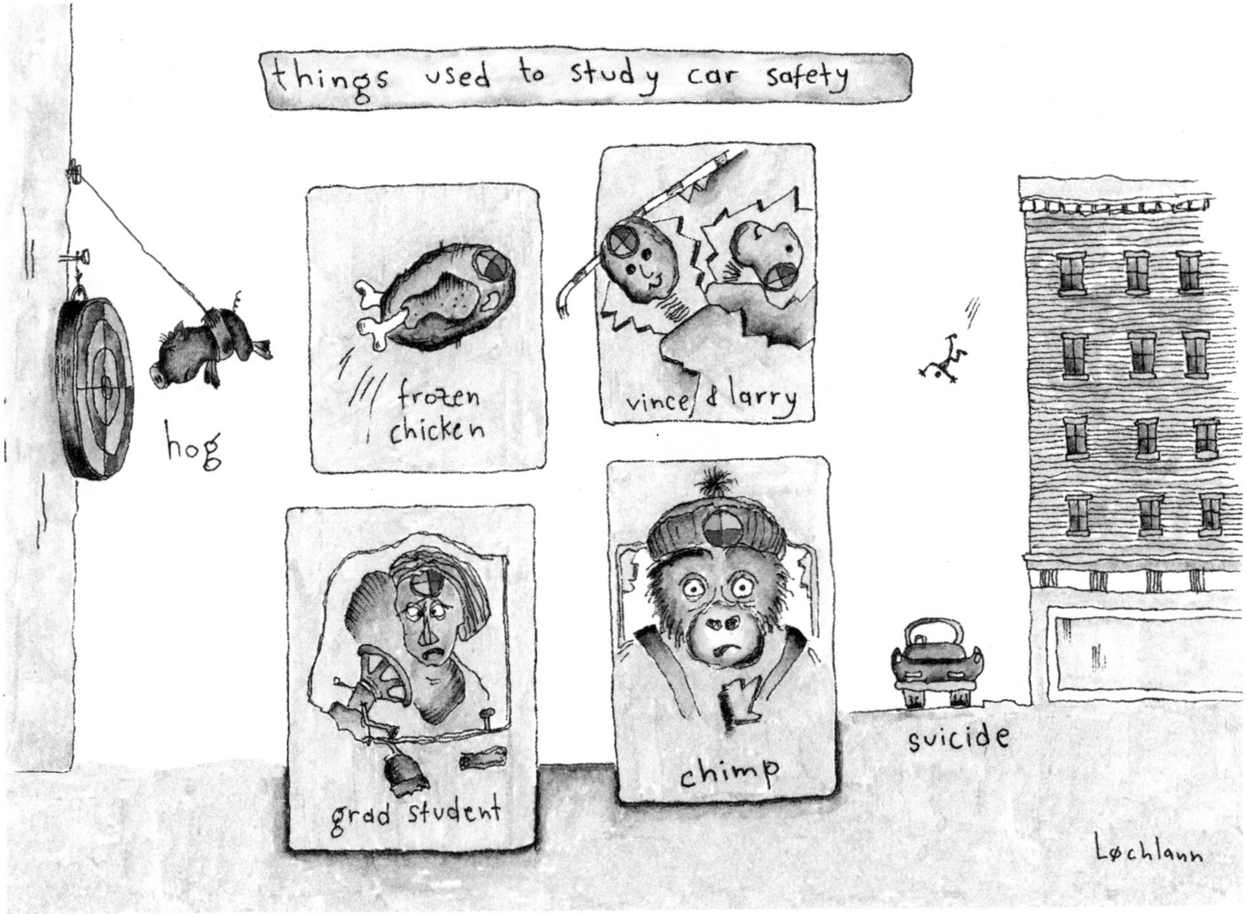
things used to study car safety
hog
frozen chicken
vince & larry
grad student
chimp
suicide
Løchlann

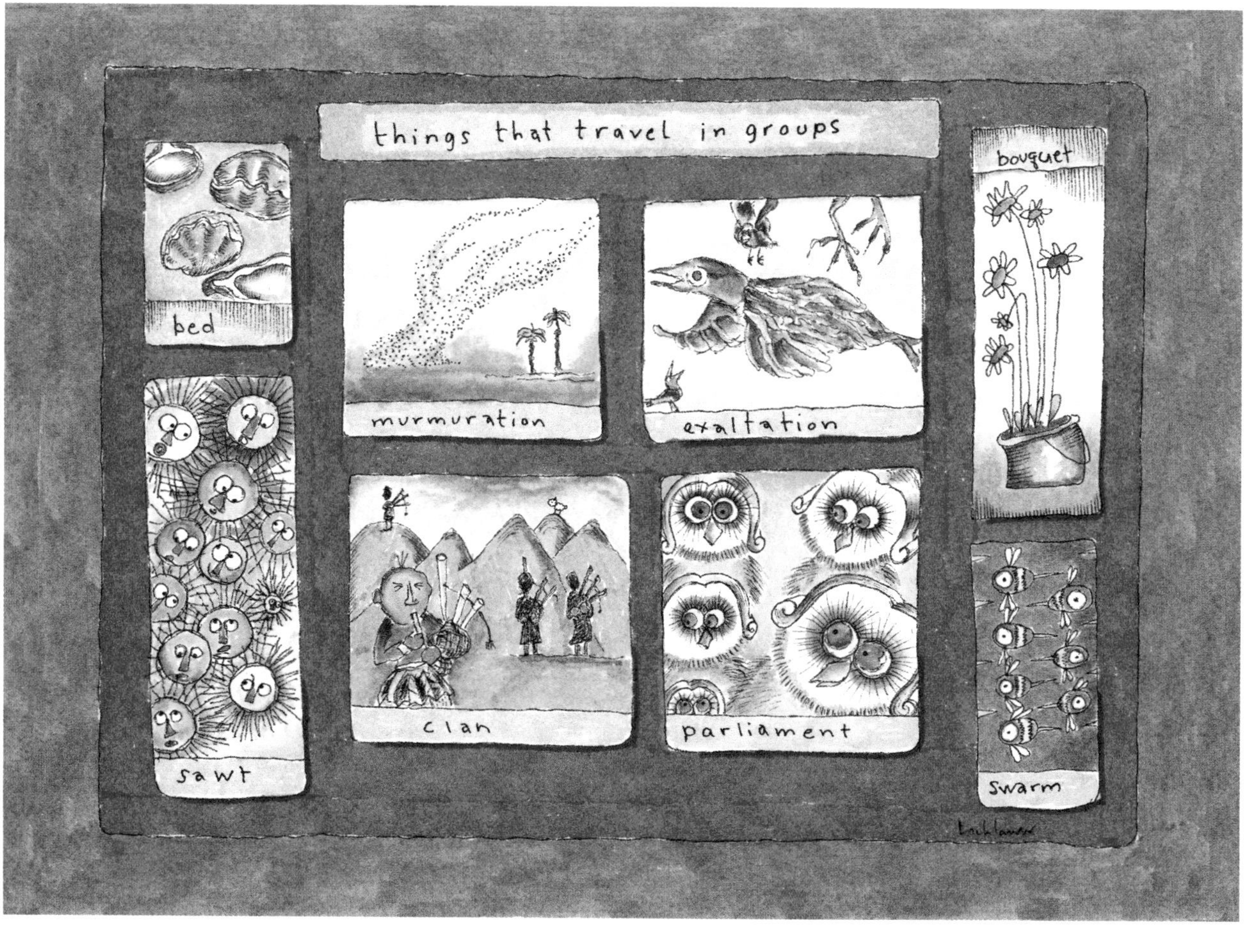

things that travel in groups
bed
murmuration
exaltation
bouquet
sawt
clan
parliament
swarm

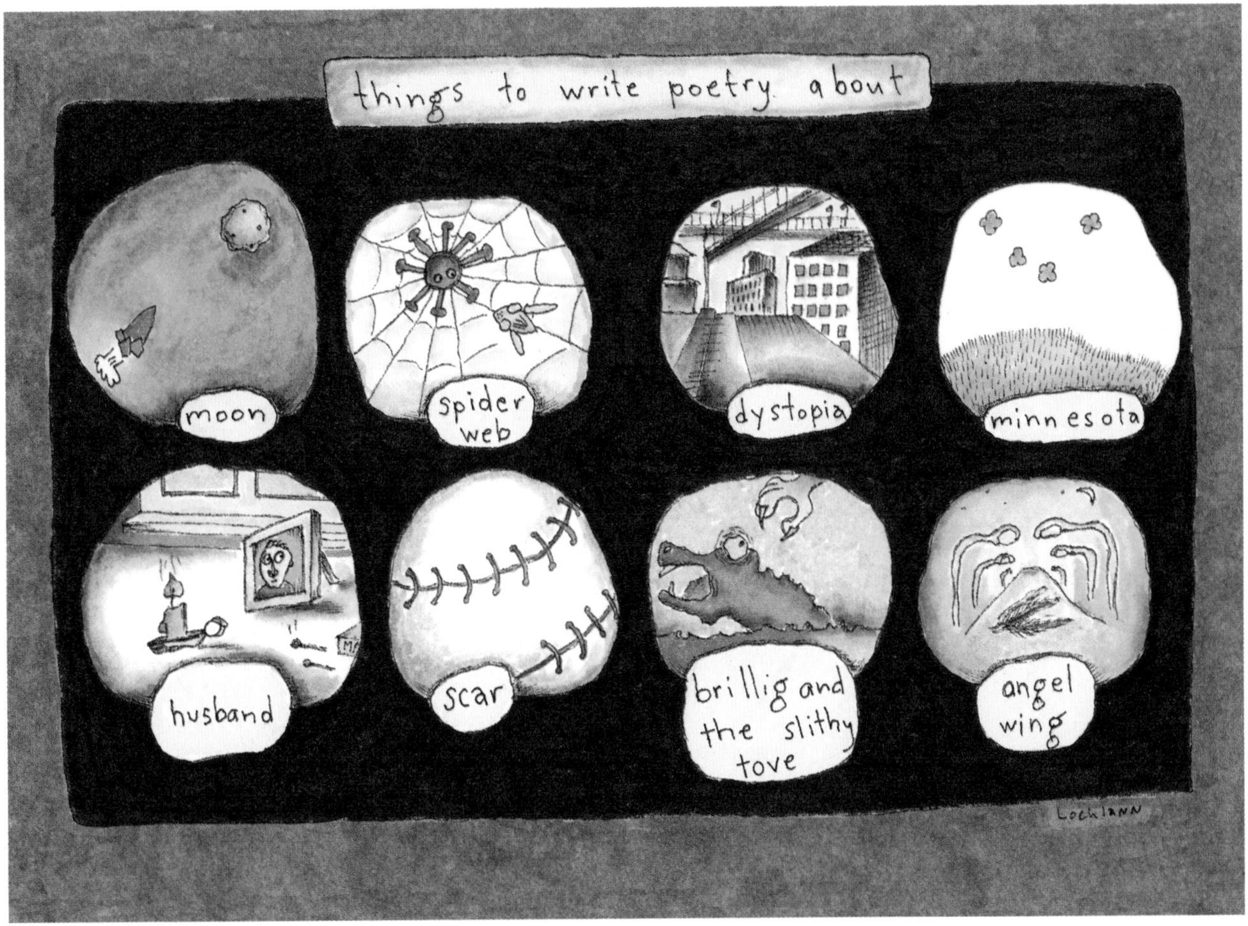
things to write poetry about
moon
spider web
dystopia
minnesota
husband
scar
brillig and the slithy tove
angel wing
LOCHLANN

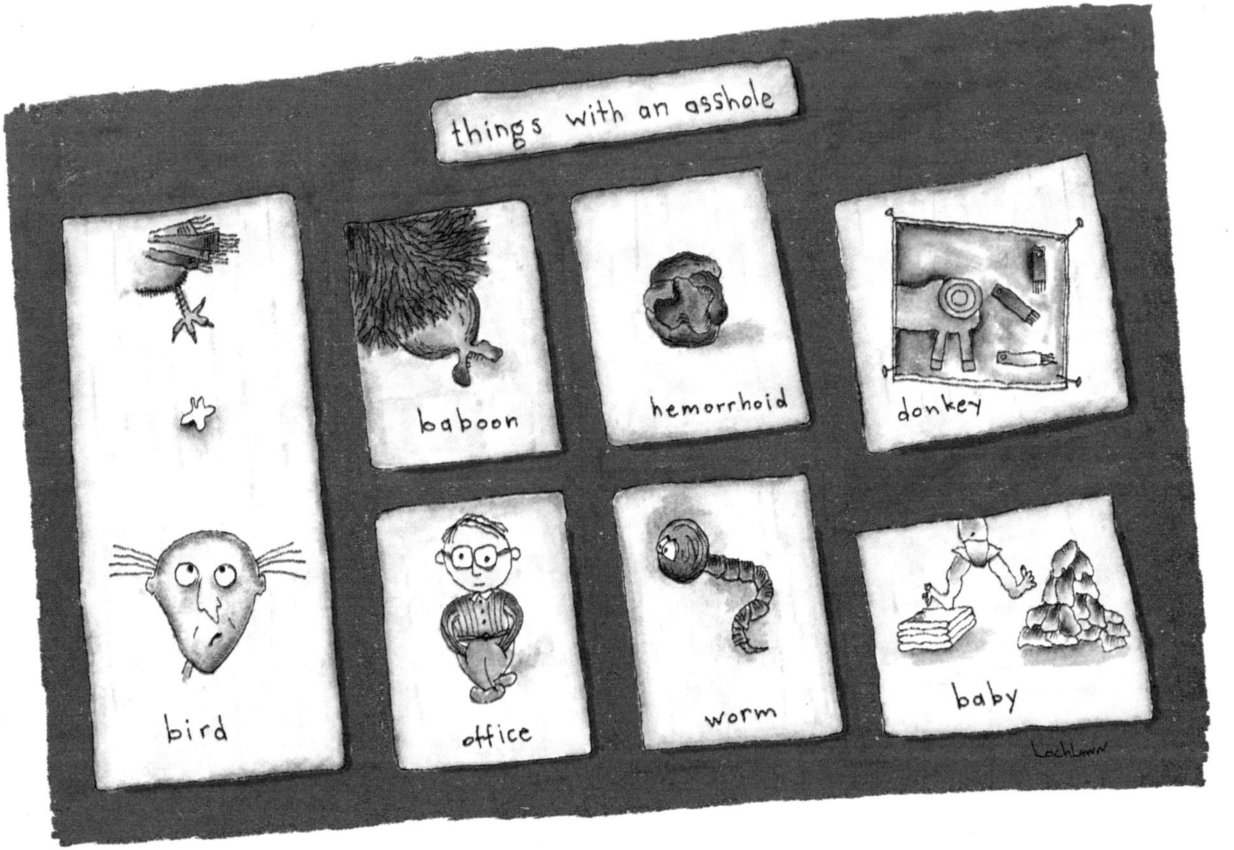
things with an asshole
bird
baboon
hemorrhoid
donkey
office
worm
baby
Lachlan

things with resolution

infection

beaker

committee

photograph

symphony

New Year's

Lochlann

things one doesn't see

cocaine

bpa
(carcinogen)

Cambodian
worker

brain

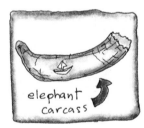

elephant
carcass

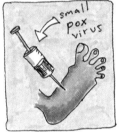

small
pox
virus

backroom
deal

school

Lochlann

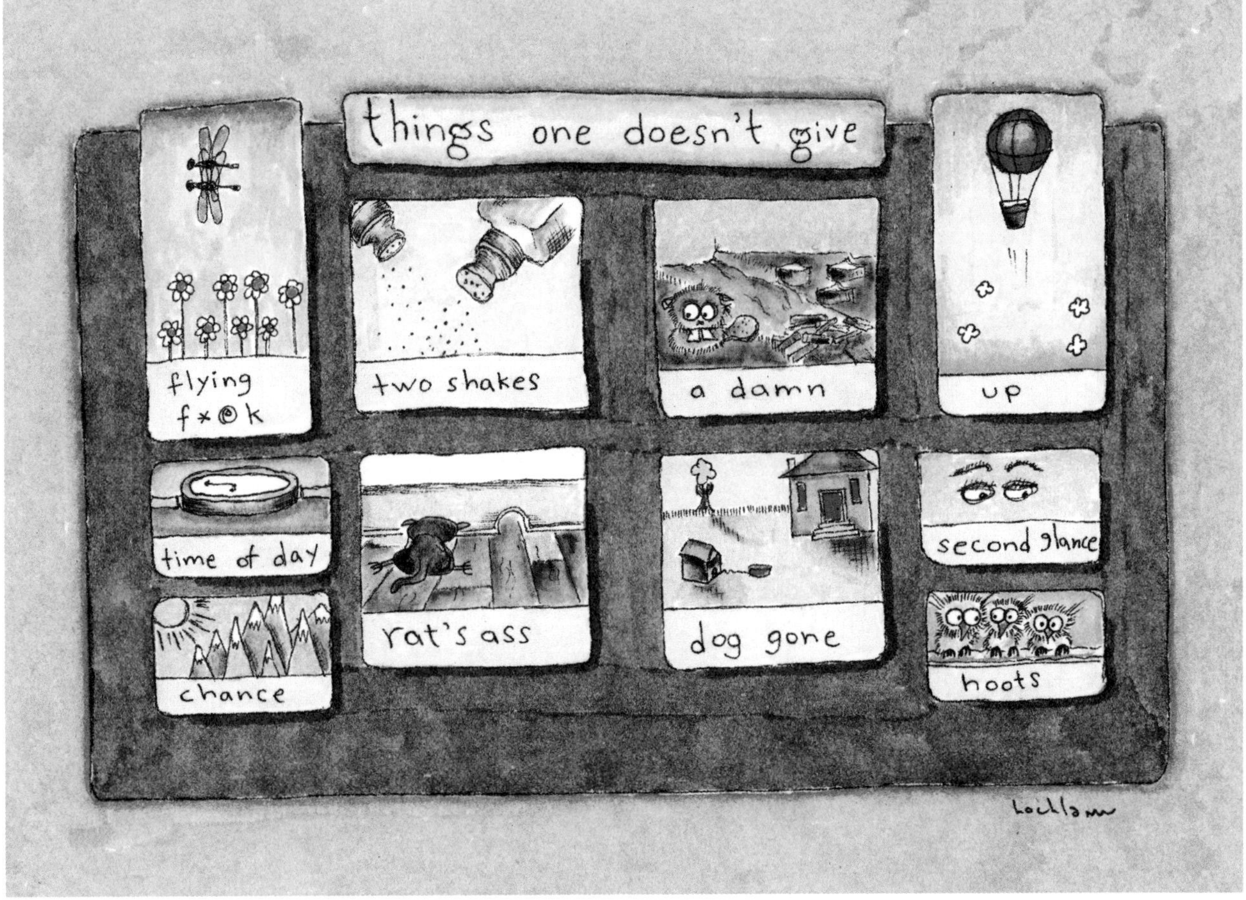
things one doesn't give
flying f*@k
two shakes
a damn
up
time of day
rat's ass
dog gone
second glance
chance
hoots
Lochlann

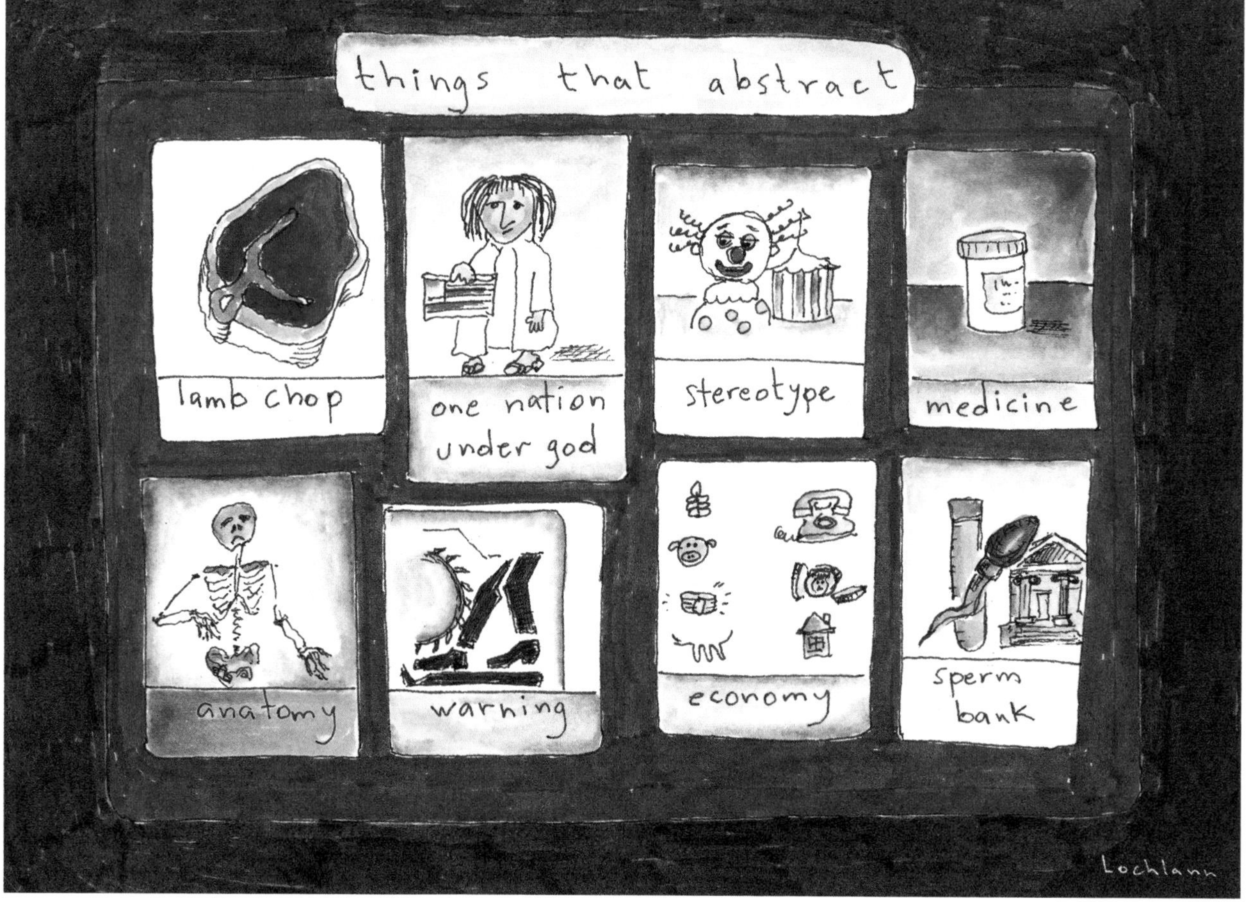

things that abstract
lamb chop
one nation under god
stereotype
medicine
anatomy
warning
economy
sperm bank
Lochlann

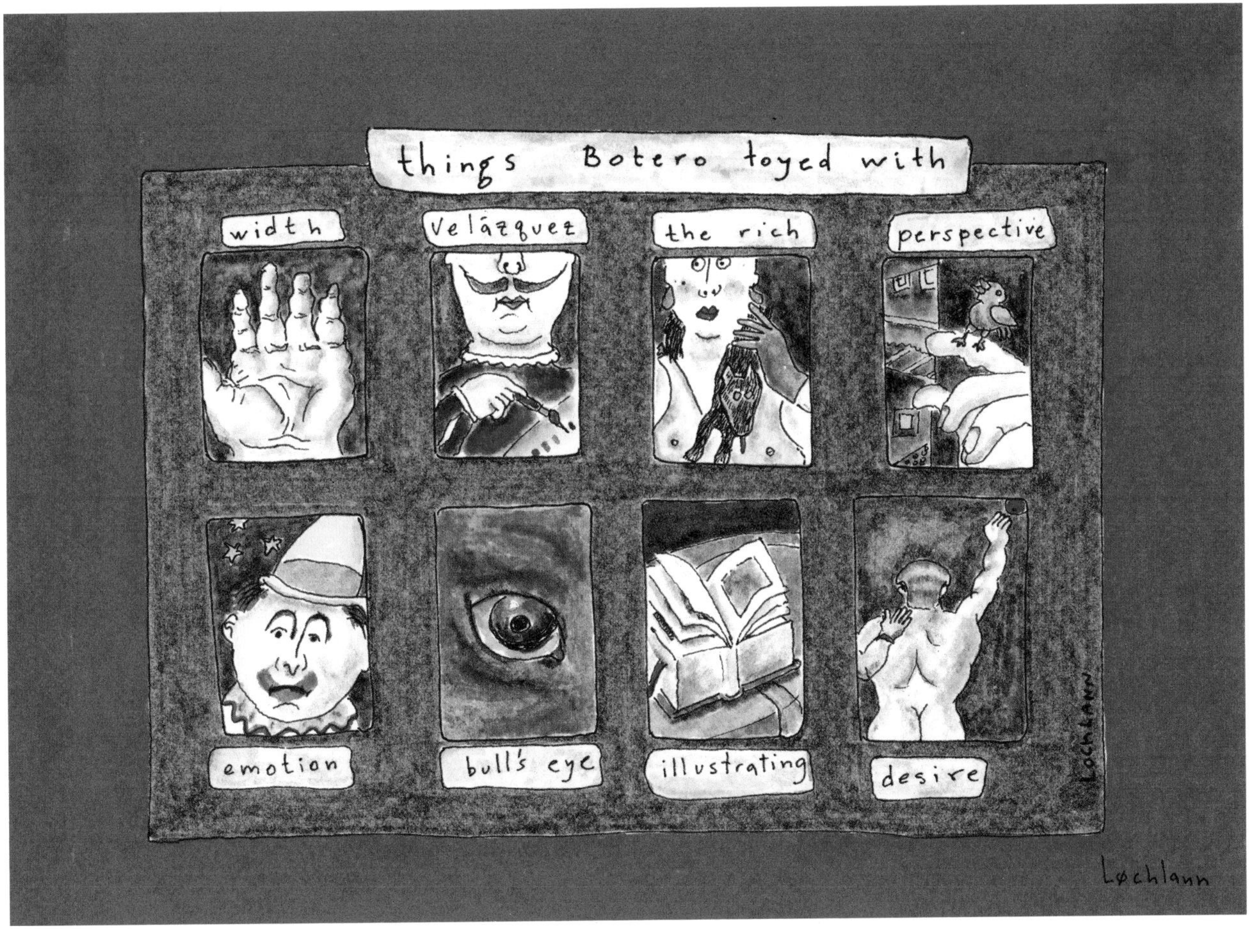

things Botero toyed with
width
Velázquez
the rich
perspective
emotion
bull's eye
illustrating
desire
Løchlann

Things That What?

DREW DANIEL

> The world is all that is the case.
> – Ludwig Wittgenstein, *Tractatus Logico-Philosophicus*

> Sea World is all that is the case.
> – Michael Robbins, *Alien vs. Predator*

So much depends upon a letter or two. With the flick of a slithering signifier, the American poet karate chops the Austrian philosopher's austere pronouncement, forcing an expansive yet precise koan to take a fishy downturn into the chum of the real.[1] It's a case in point of a general, even generic, possibility that art affords: the assault on generality itself in favor of prickly, sticky particularity. Metaphysical straight men solicit queer clowns. I think something similar is afoot with the seriously playful title of the seriously playful book you are holding in your hands. It makes your eyes do a triple take. Having expected *Things That Are*, and then *Things That Aren't*, you blink and look again at what is already before you: *Things That Art*. In this categorical assault upon categories, an anthropologist-who-is-also-an-artist makes drawings-that-are-also-arguments about the manifold, particular ways that we try, and fail, and sometimes succeed, at knowing the unruly things that make up the world.

If it were true that "the world is all that is the case," that leaves a lot of work to do, but it still sounds as if it could be done. Tally up everything that is the case and don't include anything that isn't the case and you've got the world: all the things that are and none of the things that aren't. Hope swells that a sufficiently extensive pileup of the material components of the world would give us not a model of the world but the world itself, an exhaustively inclusive Noah's Ark from which the imaginary and the non-actual have been discreetly purged. Good luck for hedgehogs and bats, bad luck for unicorns and mermaids. We already know that this can't work. The world is made up of all sorts of things, yes, but there are also processes and changes, actions and movements, and forces and flows. Things have relationships that are also a part of the world, and often it's the relationships between things, their messy processes of interaction, domination, and transformation, that make up most of what matters. From babies to tumors to viruses, new forms just keep emerging; from dodos to ice shelves to languages, old forms just keep vanishing. One could never hope to keep track of the two-way traffic between "things that are" and "things that aren't."

Accordingly, *Things That Art* insists upon the breathing room provided by a wrinkle of difference, a difference that art can make in how we imagine the ways of being available to us once we surrender the expectation that we could separate "things that are" from "things that aren't." From its title's provocation to the cabinet of curiosity within, this book insists upon the artfulness already within things, their capacity to cluster and jostle and confound our best attempts to frame and know them, whether they are as homely as "things at the farm" or as open-ended as "things that abstract." Bypassing both the imaginary black hole of non-being and the impossible enumeration of creation, in these tight, sly drawings, Lochlann Jain stages the disturbing and generative effects of things as they cling to and repel each other, and the tragicomedic role of language as it tries and fails to pin those processes down – specifically, on index cards.

Within the whirling galaxy of plural "things" on the move there's a consistent, singular thing that is always kept in view: the homely yet expansive material framework of the single index card. A stable basis for all of Jain's drawings, the index card becomes a proscenium stage on which a conceptual drama starts anew, one page at a time: a single phrase beginning with "things that ..." groups together a cluster of examples stacked in orderly rows, comprising a number of subsidiary cases. Each example is itself both a word or phrase and a tiny drawing. Jain works in miniature, with the twitchy energetic lines and sorbet color palette that, to my eyes, recalls the visual work of Roz Chast and Lynda Barry. But if this colorful and instantly accessible graphic style connotes the humble cartoon, its rendition is here cross-pollinated with the kind of persistently estranging attention to micro-implications of usage one associates with ordinary language philosophy and the fieldwork of Jain's own discipline, anthropology. Do you read these drawings or look at them? Yes.

As one works one's way across and down and sideways along Jain's grid of constituent parts, the piece's title works as both lure and frame, corralling this rebus of semiotic components into a singular-yet-composite family album. Each case contributes to the total set, but each case complicates it too, as the different timescales of "brain" and "bikini atoll" force one to consider and reconsider the invisible lines of force gathering together "things that are easily broken, slow to repair." It's only on a second or third glance that one might think of brain coral, a free associative tendril linking "brain" and "bikini atoll" together, a relationship ready to hand but lying submerged beneath the placid surface of the drawing's catalog of objects open to injury. The drawing snaps into tighter focus still when one realizes that Jain is in fact the author of an anthropological monograph on the legal and cultural nexus of injury and product design.[2]

At once describing the worn smooth pathways of our brains and building new shortcuts via tart visual/linguistic puns, Jain's associative networks condense a rich stew of reference into a beguilingly singular new form. In doing so, they resemble what Freud termed the "'collective' and 'composite figures' and the strange 'composite structures'" of dreamwork: "creations not unlike the composite animals invented by the folk-imagination of the Orient."[3] Freud's Orientalism feels like a defensive distortion of what we know he knows: the chimeras and hybrid monsters of Greek and Egyptian and Hebrew mythology closer to home. That Orientalizing riff on monstrous composite bodies summons its own composite twin in Jorge Luis Borges's *locus classicus* of taxonomic discontent, the fictitious "Celestial Emporium of Benevolent Knowledge," which divides the animal kingdom into 14 categories:

> *those that belong to the emperor, embalmed ones, those that are trained, suckling pigs, mermaids (or sirens), fabulous ones, stray dogs, those that are included in this classification, those that tremble as if they were mad, innumerable ones,*

those drawn with a very fine camel hair brush, et cetera, those that have just broken the flower vase, those that, at a distance, resemble flies.[4]

Taken together, we have a Push-Me-Pull-You monster in which authorial hand-waving toward the "Orient" becomes a thin scrim beneath which the turbulent process of ordering knowledge roils and breeds. Freud promises an etiological explanation via dreamwork that can be reverse engineered to show the component parts that make up the chimeras of the mind, while Borges reveals the specters of caprice and unreason that haunt the scene of creating the very categories into which discrete objects get placed.

Modeling this *mise-en-abyme* as if in homage to Borges's self-destroying categorical artifact, Jain's "things you chart" contains "things you chart" within itself. Borges, and Borges's influence upon Foucault, surfaces explicitly in an essay – also titled "Things That Art" – that Jain published in the journal *Anthropology and Humanism* about their practice. Speaking as both an anthropologist and an artist, Jain flags the personal stakes of their disciplinary intervention into the mesh of anthropological knowledge and its objects: "As a mixed-race, gender-fluid person, I have always had an uneasy, even antagonistic relationship to categories."[5] Identity matters, because the story of how categories work to orient and localize knowledge is itself a historical and contingent process, one that serves some interests (for Borges, the emperor; for Freud, the analyst; for Western anthropology, the university) and holds back others. In playfully constructing new categories, Jain's work models how re-zoning the freestanding categories we have inherited might dislodge their hold and make space for something else to emerge.

There's something ambitious at work here. Much of the push-pull of these drawings takes place in the tethered relationship between image and language, what Freud taught us to see as the fundamental axes of *Sachvorstellung* (thing-presentation) and *Wortvorstellung* (word-presentation) as they mesh and grind against each other to produce psychic life. Continuously probing this basic conceptual antinomy, Jain's drawings gather things on behalf of words. They help us to see language's function as a sorting mechanism, but they also draw out the latent perversity of that operation, producing comedic riffs and uncanny rebuses as a given conceptual frame opens onto vistas of application or fixates around a set of shared problems. These linguistic unions of disparate materials belong together, but the sheer arbitrary nature of how they are gathered displays the ad hoc nature of language's capacity to snag itself en route to order.

This comes to a head in "things connected by n'," because that drawing not only co-creates the assemblage it gathers, but it is also, itself, about the everyday work of assemblage making. Are the adhesions of "slip n' slide" really comparable to "mac n' cheese"? If the cuddly homosocial bonhomie of "Fred n' Barney" prompts a smile, sidling up to "Bonnie n' Clyde" pushes into the murderous terrain of *folie à deux*: whether one remembers the Peckinpah slo-mo of Bonnie and Clyde's fatal finale or not, one already senses in this drawing that the couple form isn't always quite so "soft n' cozy." To borrow a phrase from *Hamlet*, the drawing invites us to "consider too curiously" the happenstance within a seemingly trivial connection point. If "soft n' cozy" belong together through the sheer force of linguistic repetition, their link looks suddenly tendentious, subject to inversion. The drawing invites you to snuggle, but it also quietly challenges you to think of various repellent substances that are soft but far from cozy, and of cozy things (fireplaces, bed frames) that are far from soft.

There's an oddity here to Jain's encounter with the everyday, a slight but pervasive distance that is the counterpoint to its squinting proximity to the tiny. In *Light without Heat*, literary critic David Carroll Simon describes a precipitating emotional stance that has gone under-described in histories of the genesis of scientific rationalism: the dawn of a notably cool gaze upon phenomena. If that mood is one of "nonchalance" that brackets partiality as a means to try to hold its object in place, it is not without its own emotional tenor, which Simon flags in his reading of the natural philosophy of Francis Bacon as a kind of "luxurious abandon."[6] Perhaps something like this stance animates the humor that predominates in Jain's quasi-anthropological taxonomy of linguistic and material quirks, a holiday from responsibility that permits the serious work of looking at things as they are to take its own time to unfold. There's humor in this book, but it's not the sort to trigger a guffaw or a gut-bursting explosion of carnivalesque subversion. Rather, these drawings produce the quiet but precise clicking into place of a chiropractic nudge. One feels as if little adjustments are being made to one's mind as one progresses from drawing to drawing.

That sense of intuitive rightness mixed with surprise, and the relentless everydayness of its primary concerns, calls to mind an uncanny poetics of the quotidian that animates certain queer memoirs. It may just be me, but Jain's work bears a striking similarity in both tight conceptual organization and loosely paratactic feel to poet Joe Brainard's celebrated work *I Remember* (2001).[7] A list-like canticle of identical sentences, which all begin with "I remember ...," Brainard's text threads the needle of consciousness by stringing along candy bars and erections and pop songs and humdrum tasks as one continuous and seamless manifold of memory, a ticker tape of experience that darts and weaves across the tacit norms of autobiography as it hoovers up a promiscuous grab bag of examples that make up a life.

Displacing the subject or self that would organize experience, Jain starts at a different level, and it's a simple decision with crucial consequences: putting one's faith in things. This too is a modernist tactic, hearkening back to William Carlos Williams's pithy poetic slogan from his 1927 poem "Paterson": "No ideas but in things." Things were going to upstage thoughts, stand in for watery impressions with their urgent facticity and juicy, material force. To proclaim the inspirational power of "things" was to advocate for a return to the real and to charge poetry with that power. Pouncing upon the generativity that comes from starting at the level of "things," Jain's drawings take Williams at his word. But at another level, as verbal constructions, they're also not unlike poems in their compression and poise, in the exactitude of their observation and the punch of their surprise. Read straight through, "things lips do" has the zip and sting of verse. Consider how "purse/smack/stick/lick/service/fat/kiss/pucker/whistle/chap" swerves around the mouth. Neither prurient nor prudish, Jain's taxonomy of lips in motion implies all the softly powerful ways that lips can go to work on someone, including the reader in a scene of oral intimacy that both jolts and whispers. Taking compression further, "things inside things" runs ana-grammatic rings around the word's hoard of alphabetical resources, finding "sting" and "gin" and "night" and "tin" and "hi" within.

This project could go anywhere, but tendencies and fixations do emerge over the course of reading *Things That Art*. As the author of *Malignant: How*

Cancer Becomes Us, a critically celebrated analysis of the body's interanimating relationship with disease processes, Jain exhibits scholarly fixations that spill over into *Things That Art* without overly defining its exemplary range. Medical jargon bobs to the surface in "things (some) with epi," in which the Greek prefix binds together "epithelial," "epiphyte," "epidemiology," "epicenter," and "epidemic" into an ominous garland, and Jain's familiarity with the social history of medicine returns in the causal insertion of "cholera" into "things you chart." For those with the eyes to see, the microscopic drawing quietly evokes Dr. John Snow's celebrated mapping of a London cholera outbreak to a particular water pump on Soho's Broad Street, and, by extension, the possibility of charts to not only describe the world but intervene in and upon it, in this case for the public good.[8] Alert as they are to how categories harm and distort, Jain reckons with their occasionally lifesaving potentialities, too.

Is there any limit to what we can find within "things"? Language tricks us into thinking that the singularity of a verbal descriptor will produce a limited field of reference, but when the word in question is "things," the flood of instances keeps widening and expanding. Worrying at the limits of their own commitment, Jain asks: "How to include negatives and futures within linguistic conventions of things? If a noun is a person, place or thing – as every school child knows – is a thing always necessarily a noun?"[9] In pursuit of the messy processes and unstable motion within things, Jain activates the word. This leaps off the page in "things a dollar does," which turns properties into actions and spins currency off axis. The capacity of the money form to accumulate and crash is belied by its status as all-too-fragile matter: a dollar may buy, but it can also "float" and "burn." If, as Sara Ahmed puts it, "objects bring worlds with them,"[10] the object of a dollar here stands in for capitalism's world-building force, but it also reminds us of a material surround that stands outside and beyond the economy. In that pointed reminder of the capacity of a dollar to float and burn, Jain forces us to remember our own purchase – so to speak – upon the world. To state the obvious, it matters whether we are a philosopher in a villa in Vienna or a dolphin caged in a tank in Florida: Who has what range of motion, and why? From "things at the farm" to "things that abstract," the world of objects is a world of both constraint and possibility, a world made up of things and the frames for things, an assemblage of ongoing categorical work of inclusion and exclusion, containment and release. Sidestepping the distinction between things that are and things that aren't, Jain reminds us that a shared world remains open to change, or, to use another word, to art.

Notes

1 See Ludwig Wittgenstein, *Tractatus Logico-Philosophicus*, trans. C.K. Ogden (London: Routledge and Kegan Paul, 1981) and Michael Robbins, "Downward Facing Dog," in *Alien vs. Predator* (New York: Penguin, 2012).
2 S. Lochlann Jain, *Injury: The Politics of Product Design and Safety Law in the United States* (Princeton: Princeton University Press, 2006).
3 Sigmund Freud, "On Dreams," in *The Freud Reader*, ed. Peter Gay (New York: W.W. Norton, 1989), 153.
4 Jorge Luis Borges, "John Wilkins' Analytical Language," in *The Total Library: Non-Fiction 1922–1986*, ed. Eliot Weinberger, trans. Esther Allen, Suzanne Jill Levine, and Eliot Weinberger (London: Penguin, 2001), 229.
5 S. Lochlann Jain, "Things That Art," *Anthropology and Humanism* 43, no. 1 (2018): 6, https://doi.org/10.1111/anhu.12198.

6 David Carroll Simon, *Light without Heat: The Observational Mood from Bacon to Milton* (Ithaca: Cornell University Press, 2018), 5.

7 Joe Brainard, *I Remember* (New York: Granary Books, 2001).

8 For more on Snow's map, see Steven Johnson's *The Ghost Map: The Story of London's Most Terrifying Epidemic – and How It Changed Science, Cities, and the Modern World* (New York: Riverhead Books, 2007).

9 Jain, "Things That Art," 10.

10 Sara Ahmed, *Living a Feminist Life* (Chapel Hill: Duke University Press, 2014), 41.

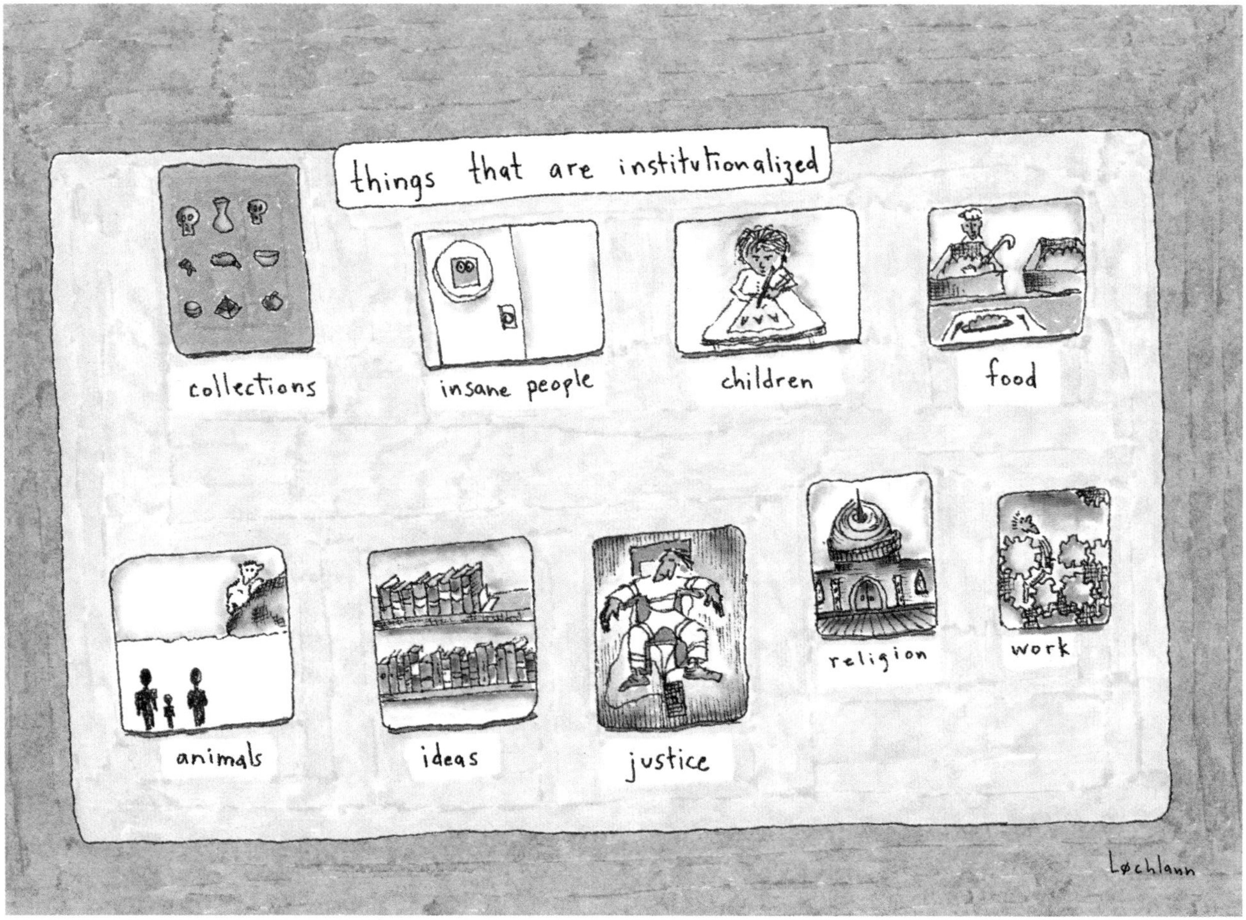
things that are institutionalized
collections
insane people
children
food
animals
ideas
justice
religion
work
Løchlann

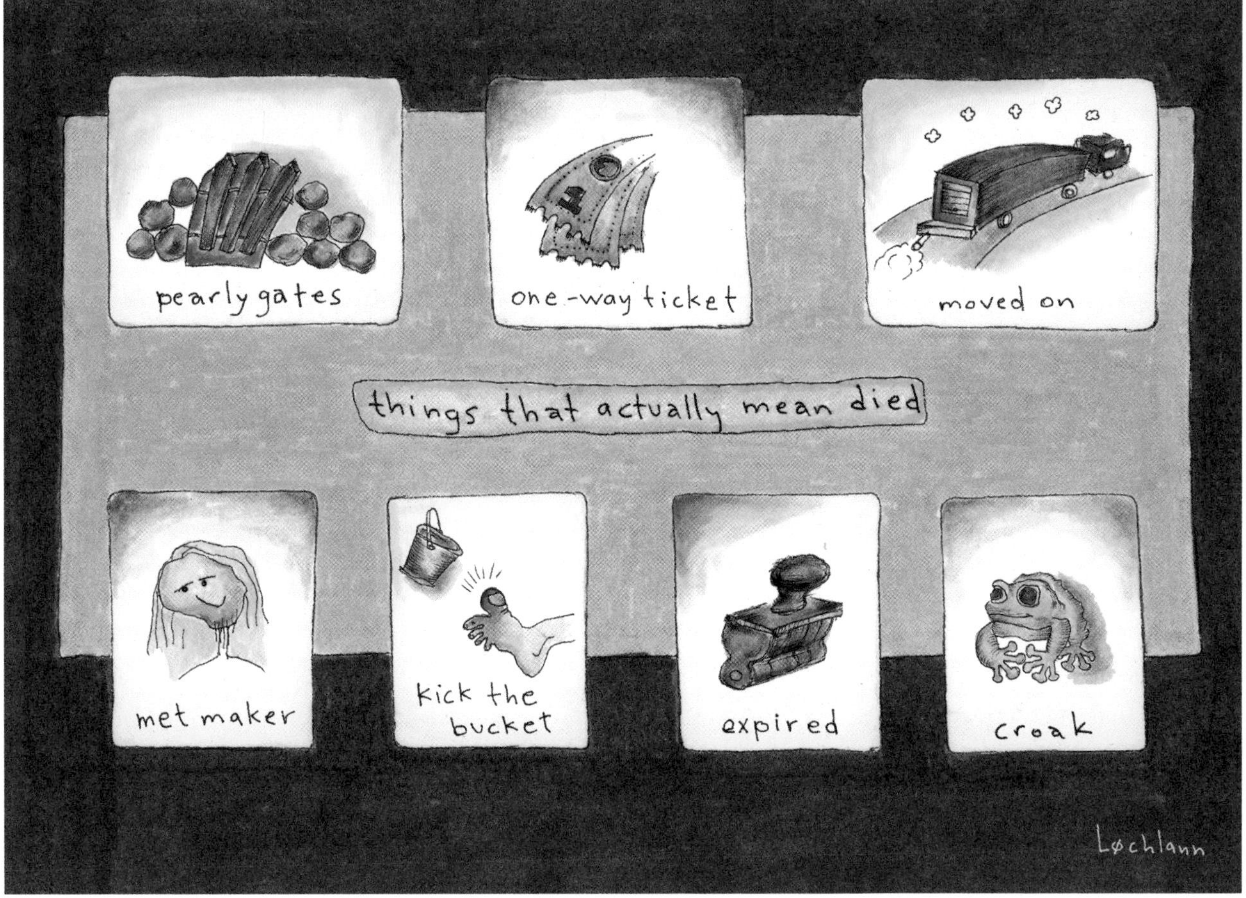
pearly gates
one-way ticket
moved on
things that actually mean died
met maker
kick the bucket
expired
croak
Løchlann

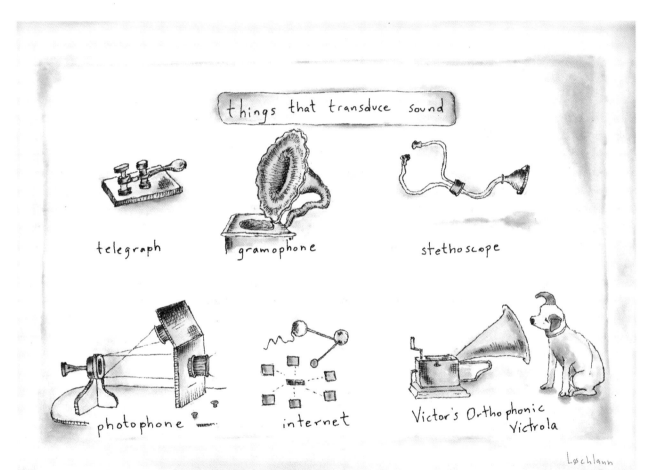

things that transduce sound

telegraph

gramophone

stethoscope

photophone

internet

Victor's Orthophonic Victrola

Lochlann

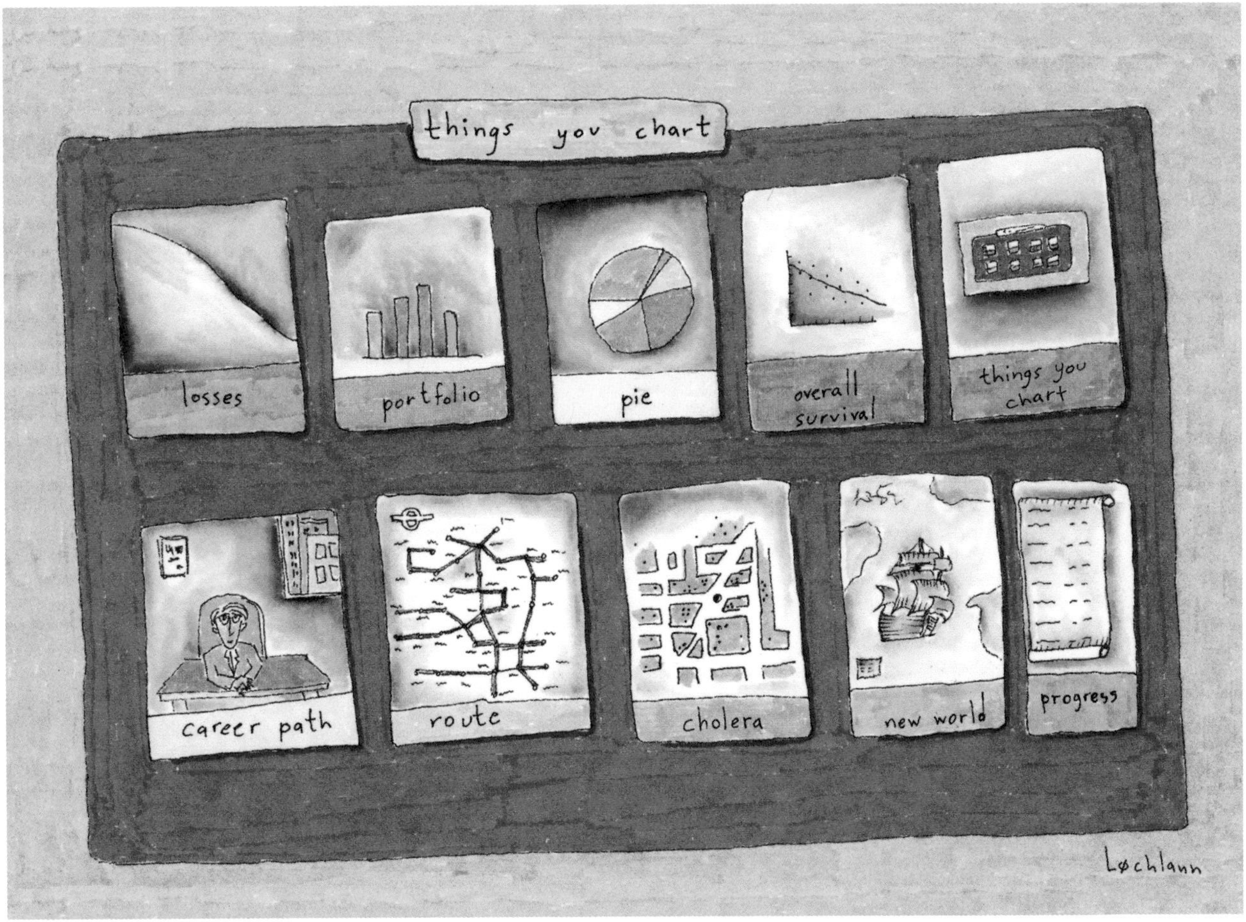
things you chart
losses
portfolio
pie
overall survival
things you chart
career path
route
cholera
new world
progress
Løchlann

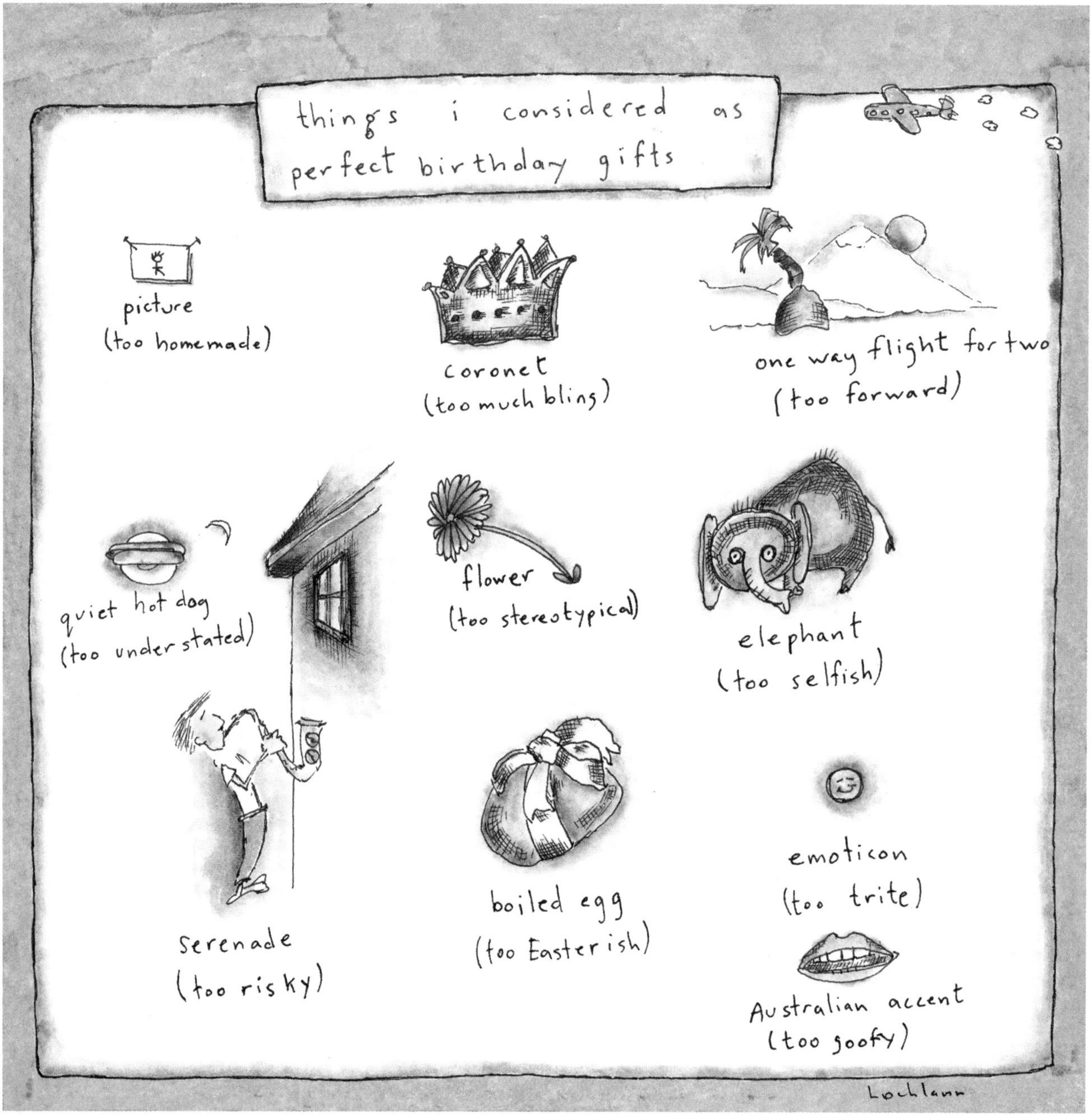
things i considered as perfect birthday gifts
picture
(too homemade)
coronet
(too much bling)
one way flight for two
(too forward)
quiet hot dog
(too understated)
flower
(too stereotypical)
elephant
(too selfish)
serenade
(too risky)
boiled egg
(too Easterish)
emoticon
(too trite)
Australian accent
(too goofy)
Lochlann

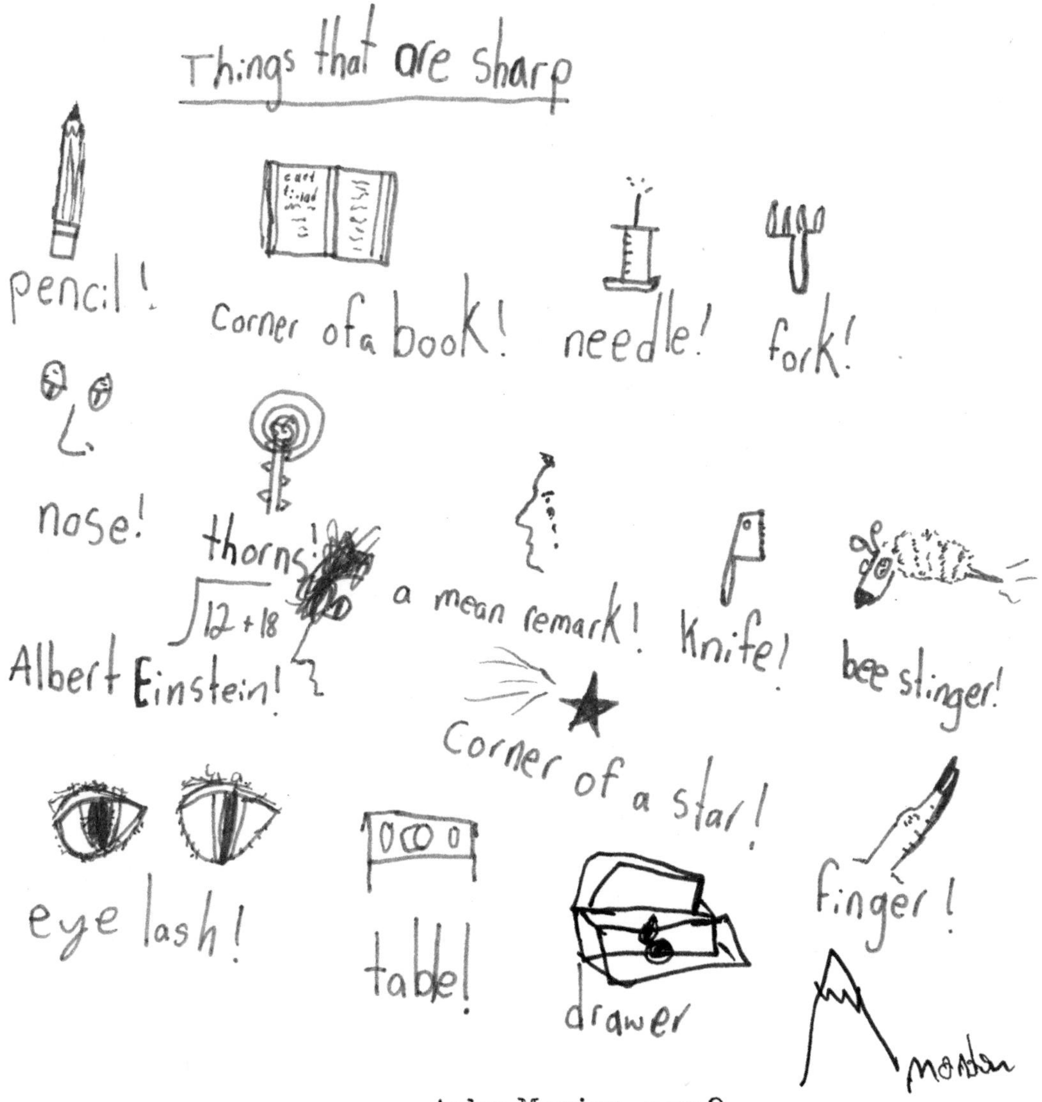
Things that are sharp
pencil!
corner of a book!
needle!
fork!
nose!
thorns!
12 + 18
Albert Einstein!
a mean remark!
knife!
bee stinger!
Corner of a star!
eye lash!
table!
drawer
finger!

Asha Marion, age 9

things that are not

concrete

fair

$$\sqrt{6} \times 9y = .9^4 7\mu$$
$$48\theta zy \geq 4\zeta$$

well known

black & white

alive

nothing

safe

the sharpest pin

perfect

Løchlann

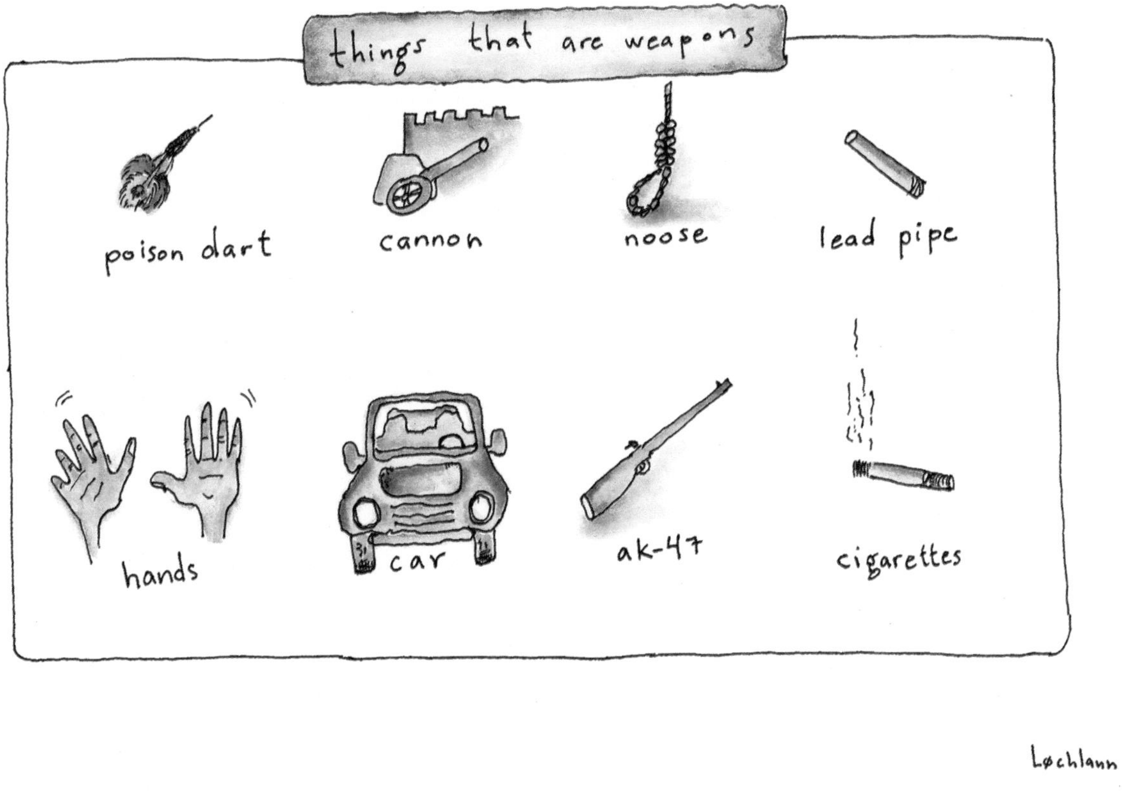
things that are weapons
poison dart
cannon
noose
lead pipe
hands
car
ak-47
cigarettes
Løchlann

things generally mentioned as insults
bastard
swine
dickhead
motherfucker
square
fag
princess
creep
bum
cheapskate
dyke
bitch
Løchlann

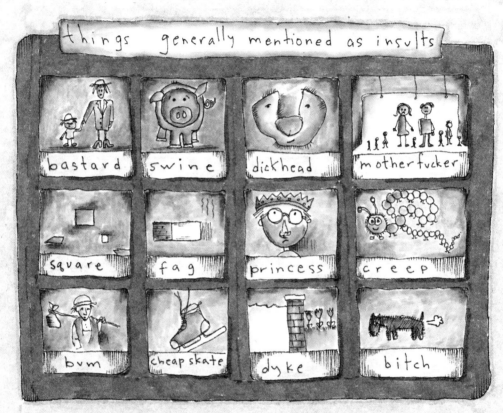

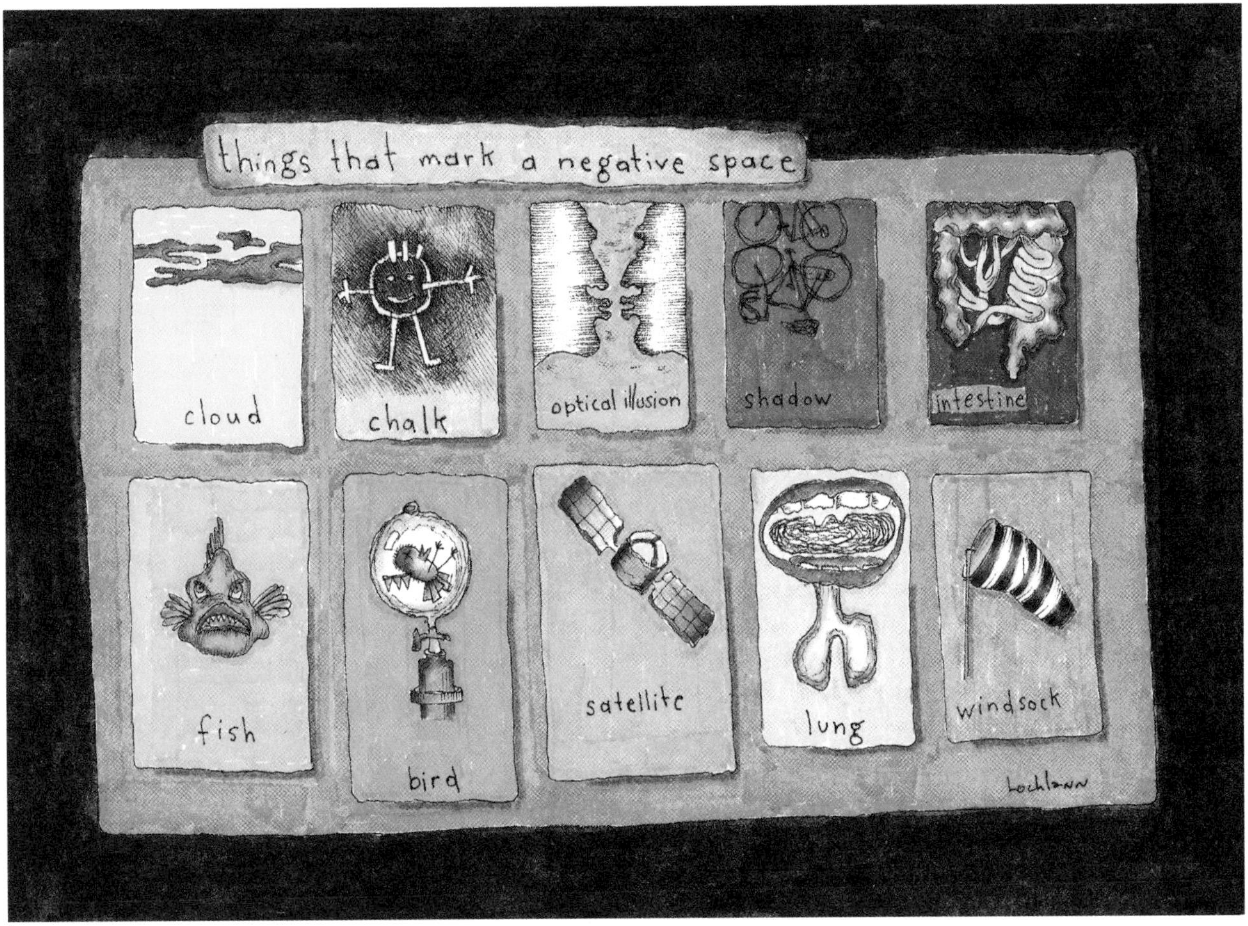
things that mark a negative space
cloud
chalk
optical illusion
shadow
intestine
fish
bird
satellite
lung
windsock
Lochlann

things that are visible and not

gold

light

water

category

phosgene

beauty

love

time

Løchlann

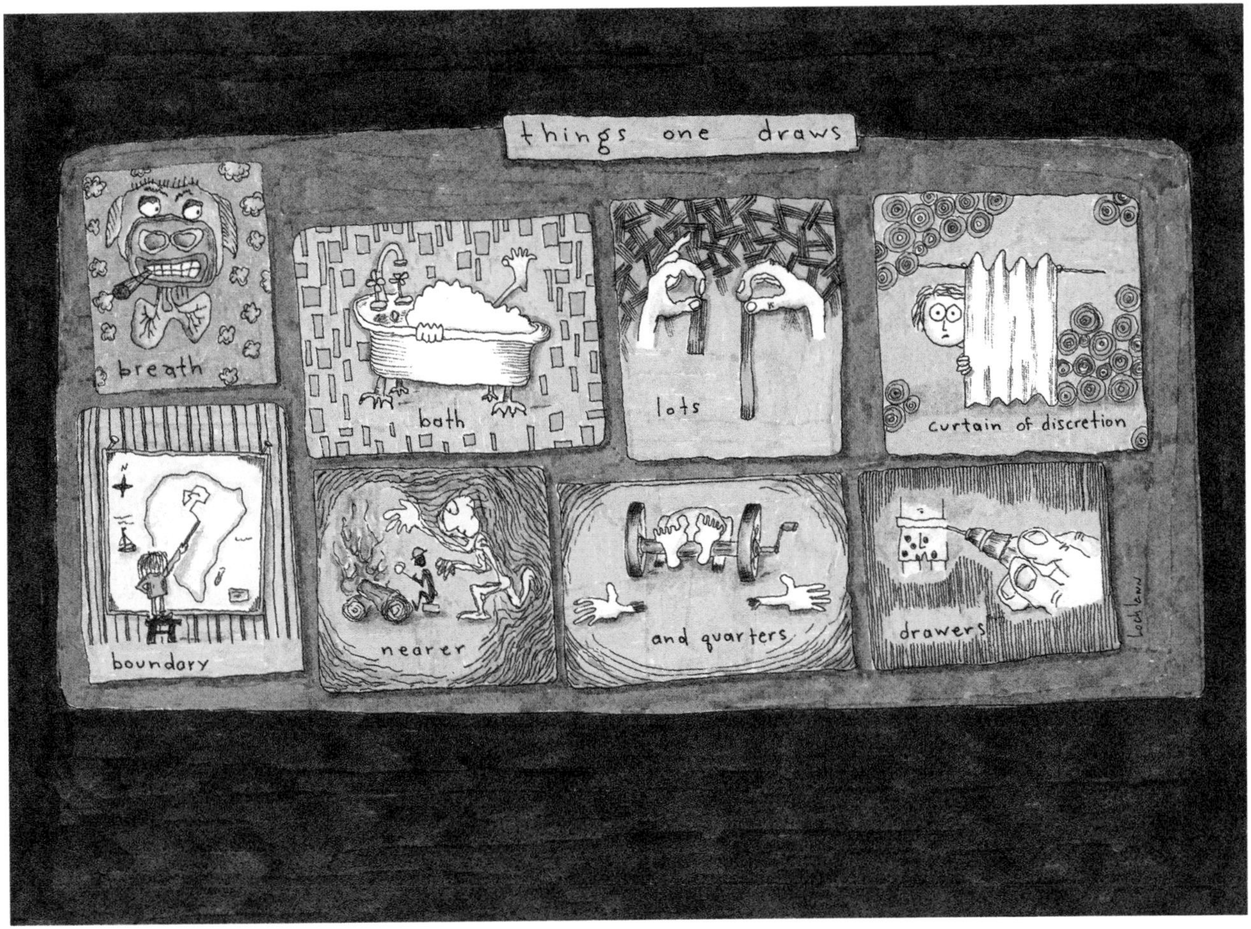
things one draws
breath
bath
lots
curtain of discretion
boundary
nearer
and quarters
drawers

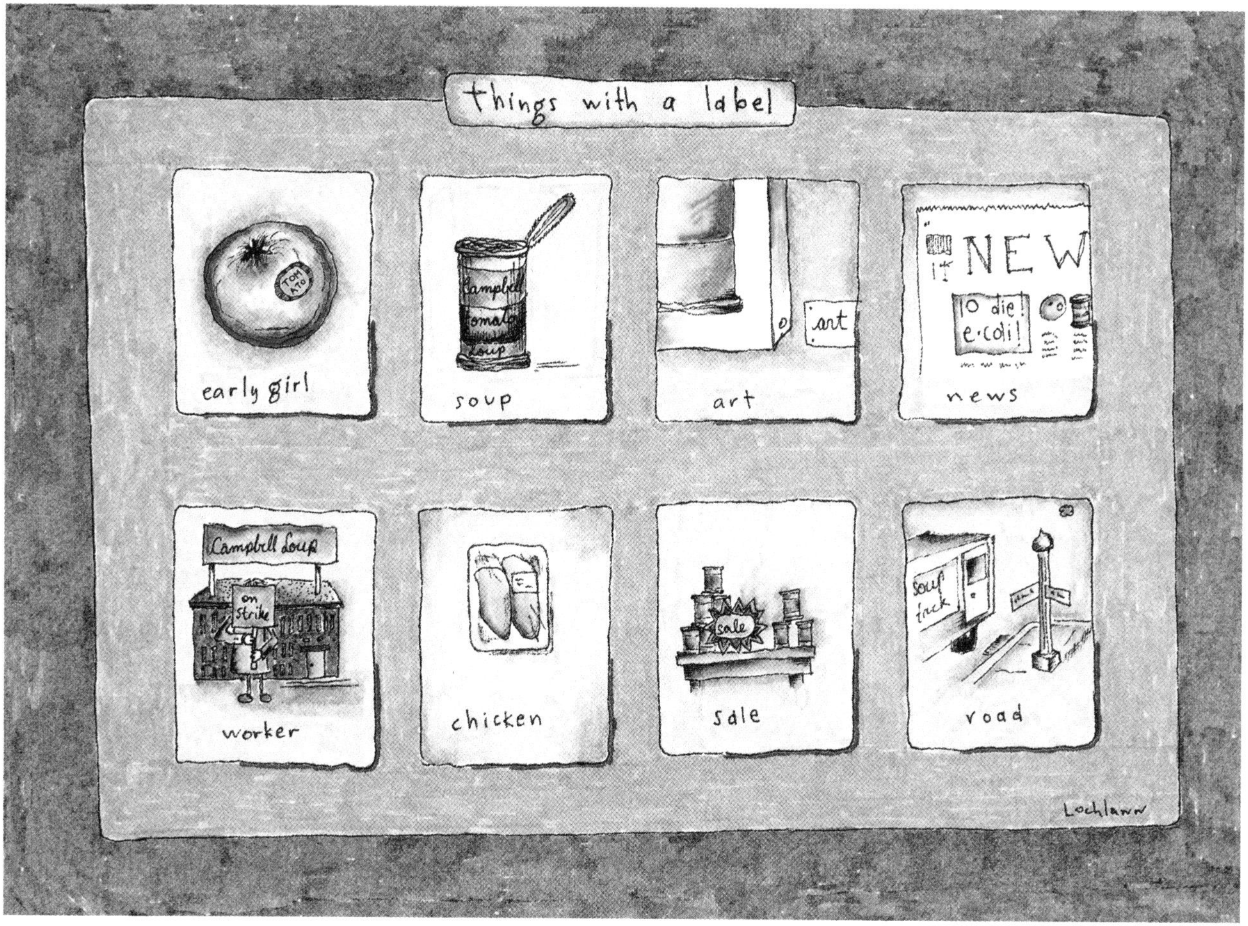
things with a label
TOMATO
early girl
Campbell
Tomato
Soup
soup
art
art
NEW
10 die!
e.coli!
news
Campbell Soup
on strike
worker
chicken
sale
sale
soup truck
road
Lochlann

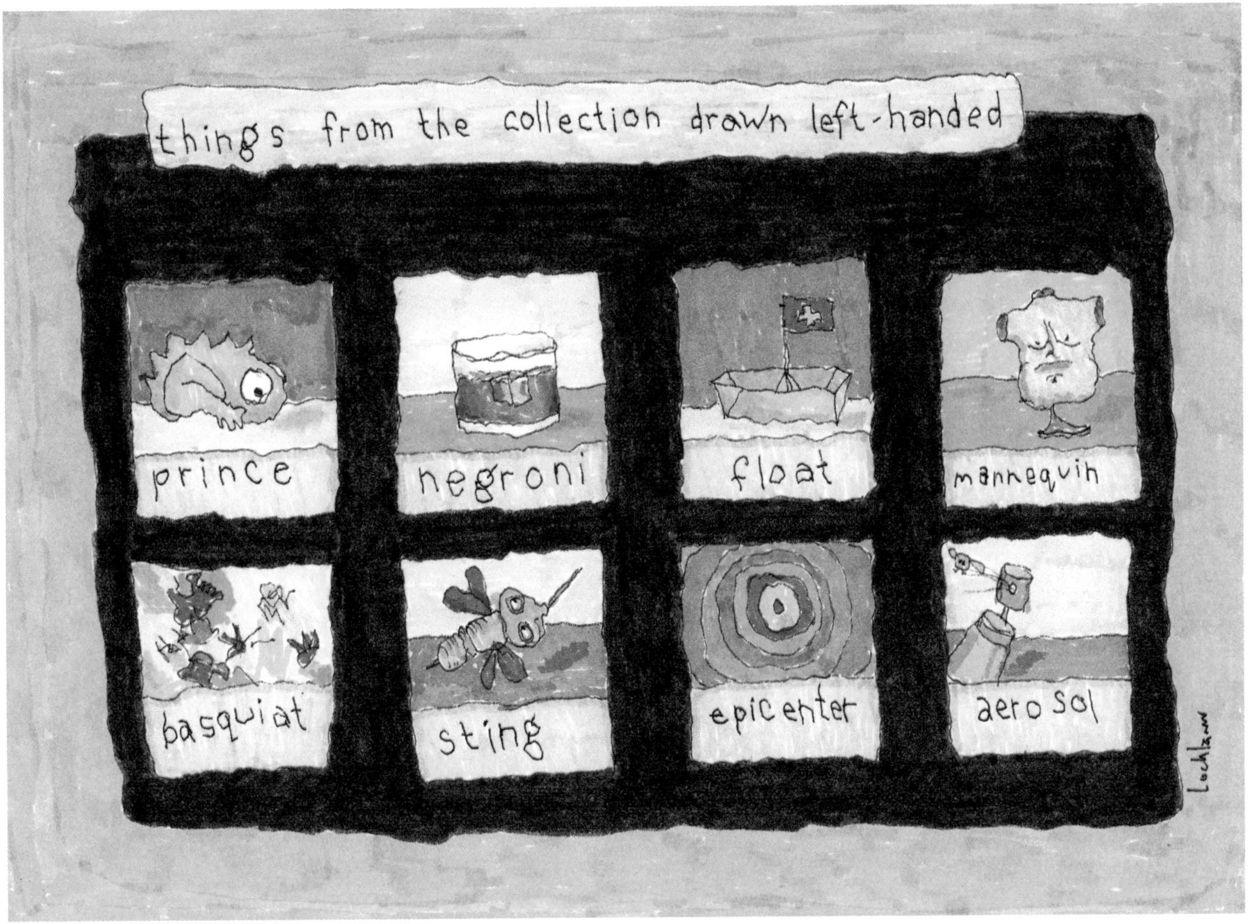
things from the collection drawn left-handed
prince
negroni
float
mannequin
basquiat
sting
epicenter
aero sol
Lochlann

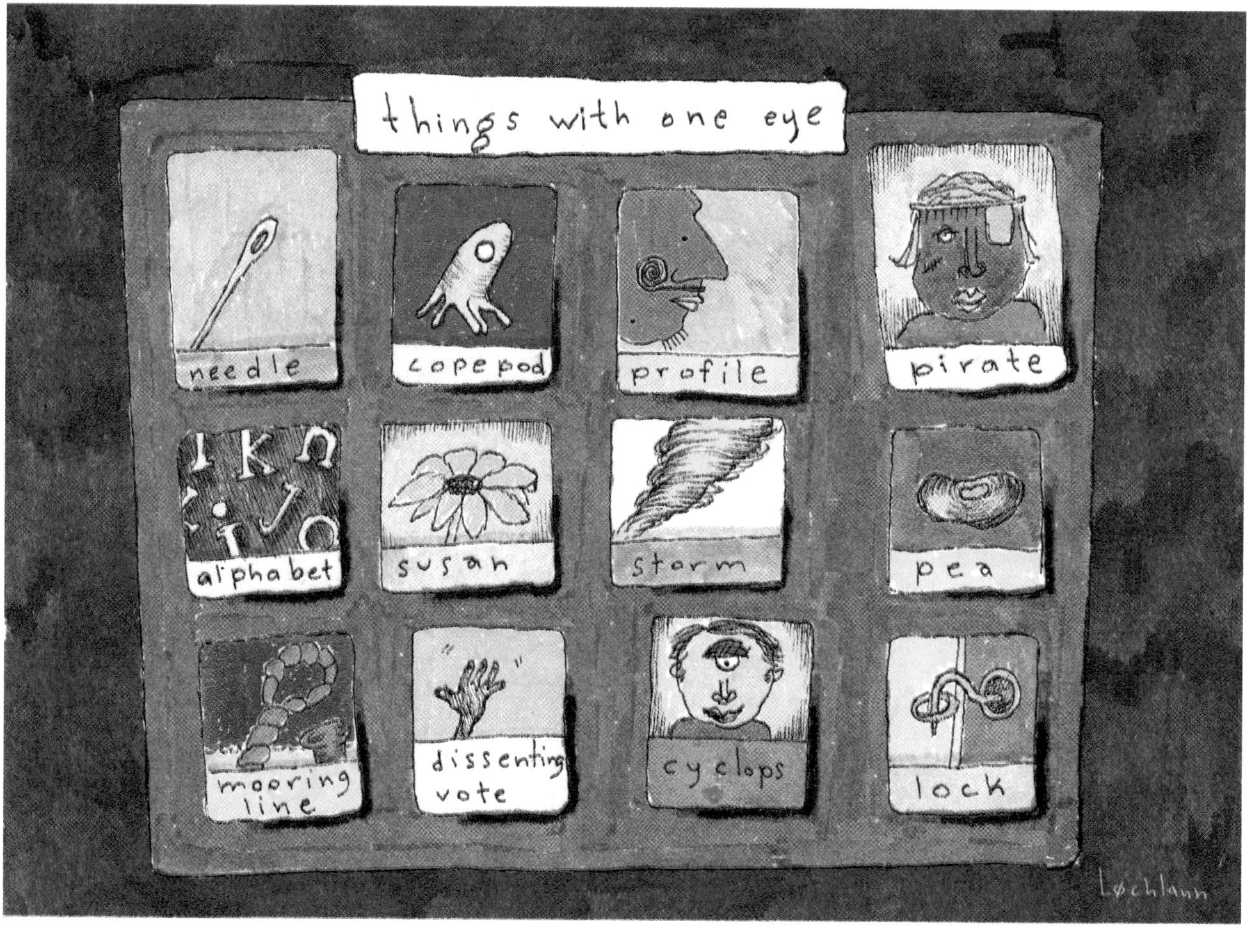
things with one eye
needle
copepod
profile
pirate
alphabet
susan
storm
pea
mooring line
dissenting vote
cyclops
lock
Løchlann

things some people jolly well eat but we tend not to

marrow

horse

lard

cricket

kangaroo

hat

knuckle sandwich

spotted dick

crow

tongue

nothing

dust

Lochlann

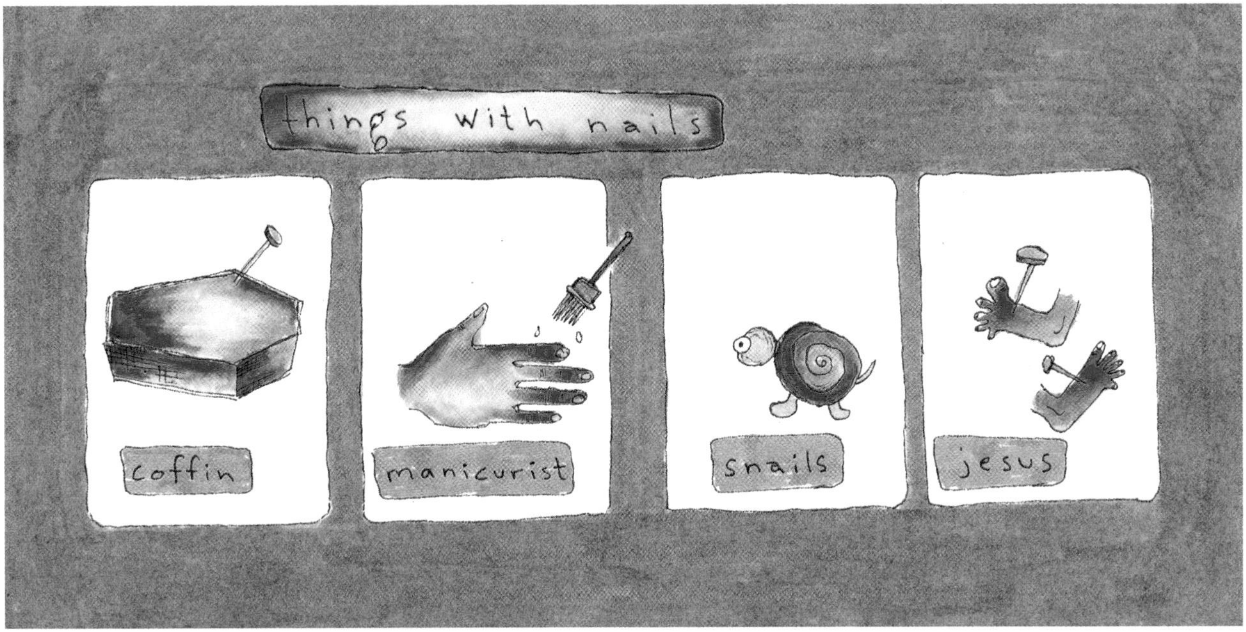
things with nails
coffin
manicurist
snails
jesus

Løchlann

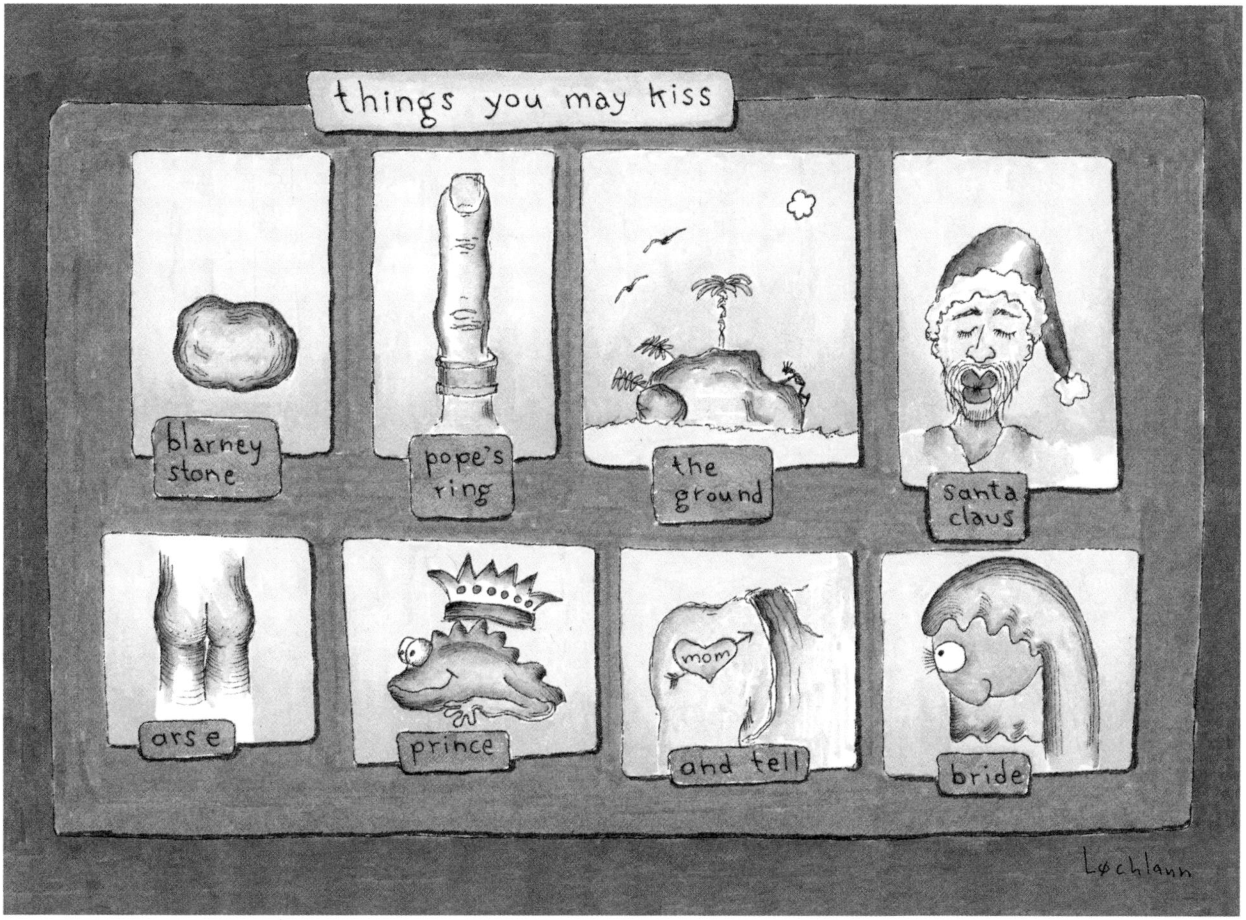
things you may kiss
blarney stone
pope's ring
the ground
santa claus
arse
prince
mom
and tell
bride
Løchlann

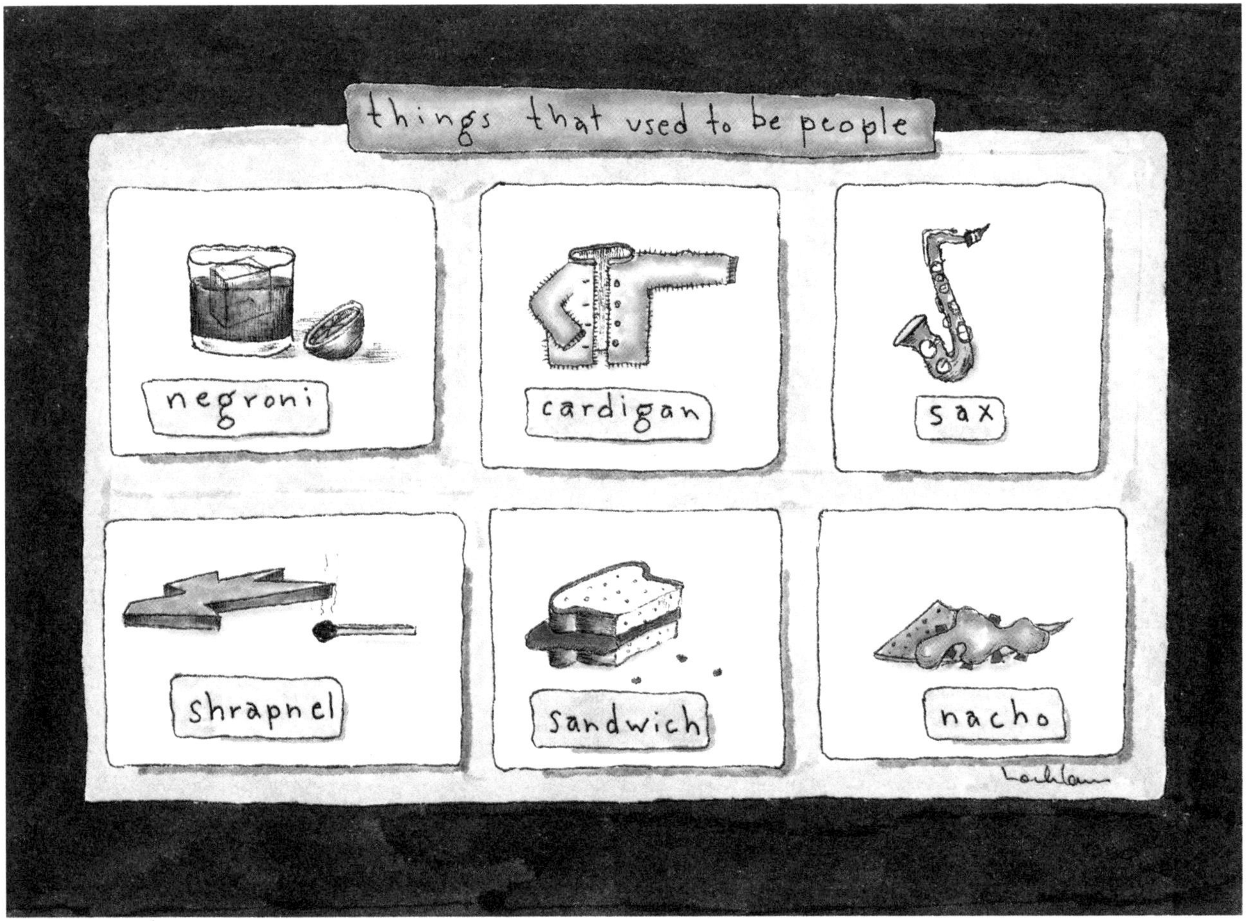
things that used to be people
negroni
cardigan
sax
shrapnel
sandwich
nacho

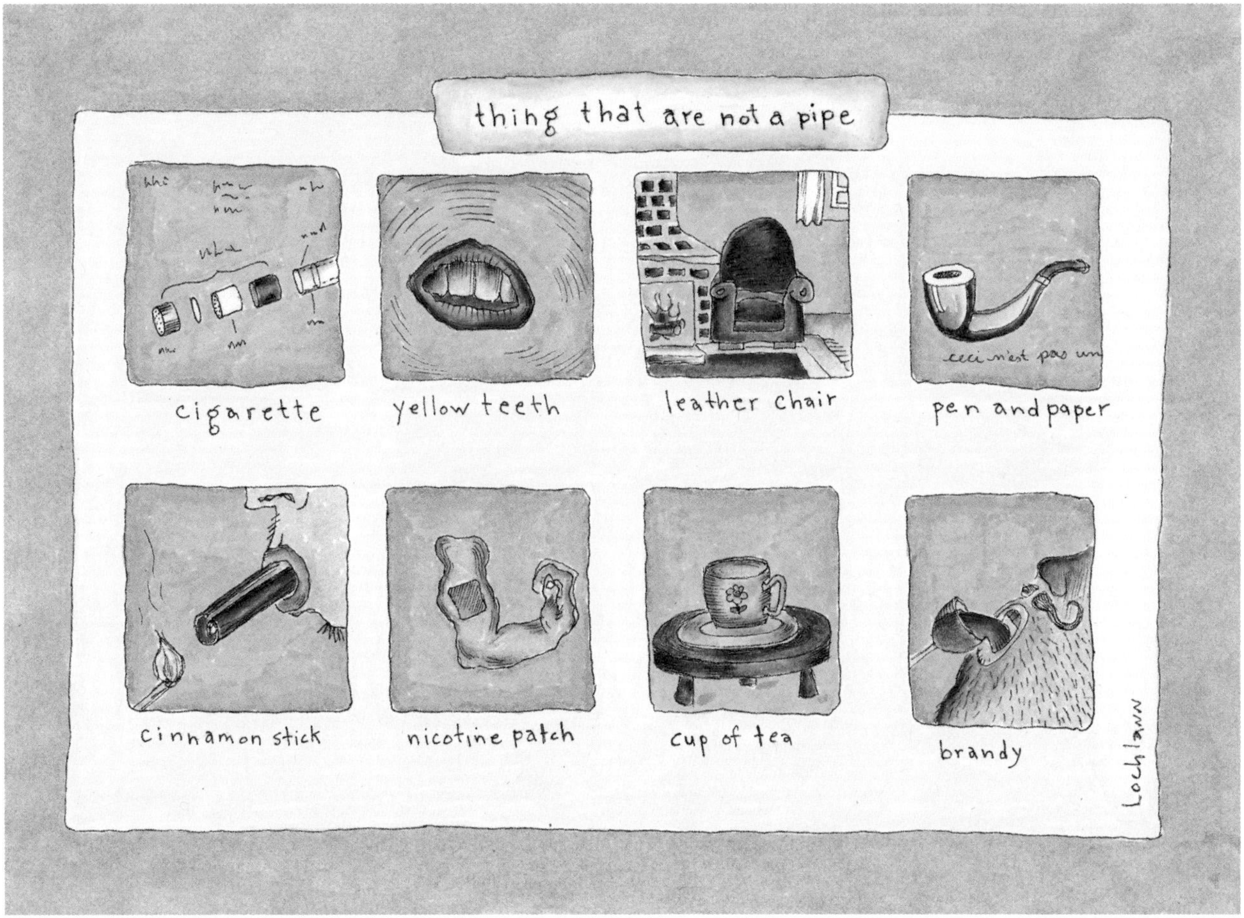
thing that are not a pipe
ceci n'est pas un
cigarette
yellow teeth
leather chair
pen and paper
cinnamon stick
nicotine patch
cup of tea
brandy
Lochlann

Which Things Mean When

LOCHLANN JAIN

Savants throughout history have assumed that in an ideal and dependable world all things would, like puzzle pieces and children, stick to their designated spots. Aristotle divvied the world into unadorned, albeit fervent, clusters: quantity, quality, and passion. Eighteenth-century botanist Carl Linnaeus preferred spooning: kingdom curls around genus whorls species.

Linnaeus's stockpile of index cards, inscribed with the names of plants and animals, outperformed even Silicon Valley's legendary cocktail napkins on which so many start-up ideas have been sketched. The humble paper scraps offered shorthand for the bodies and souls of grizzlies and damselflies, rendering them at once mobile and comparable, although risking the loss of key individual features. Linnaeus sorted and pasted his way to a system that annotated *Tyrannosaurus rex* as well as it did *Homo sapiens*. Card, category, label, collection: invented technologies that at once enable and eclipse.

Linnaeus's *Systema Naturae* (1735) coincided with a period of European mercantile interests vying to bring home the most extraordinary, incomprehensible, and – as often as not – stolen exoticisms, from tea and typhus to coffee and cocaine. Still, it would be a mistake to merge the distinct projects of colonializing, collecting, and categorizing. The adequacy of words and images to fully render things has long been suspect. (One might recall the 1:1 map described by Lewis Carroll in *Sylvie and Bruno Concluded* [1893]. The farmers objected to spreading out the unqualified, unabbreviated, all-inclusive representation on the grounds that it blocked the sun.)

Sylvia Pankhurst, a scholar widely known as a suffragette, describes equally fantastical if more sober efforts to spackle these inherent fissures between the world and representations of it. A priori languages start afresh, seeking to design systems that align the sounds and scripts of language with the actual things they aim to describe.[1] One such endeavor offers both a fascinating precedent to Linnaeus and an insight into world-constituting recipes.

If Linnaeus wisely limited his scope to plants and critters, John Wilkins, the first secretary of the Royal Society of London, set his sights on nothing less than the great chain of being. *An Essay towards a Real Character, and a Philosophical Language* (1668) describes an entire, and entirely new, language, one that aimed to obliterate the distinction between word and thing – each noun encoding the full set of relations in which it was embedded.

Wilkins's baroque language remarkably portends Linnaeus's nested hierarchies. For example, in one of his many diagrammed concept-relations, one finds a pedigree for "hanging." Organized as if on a family tree, a list of judicial terms appears thus: *judicial relation → punishments → capital → simple → separation of the parts → interception of the air → [at the] throat → hanging*. If that weren't poetic enough, Wilkins conjures a multi-phonic onomatopoeia as the basis for his language such that to know the *word-formerly-known-as-hanging* would be also to grasp the stock and stake of the concept within this set of relations.

Living languages learn, and Wilkins doesn't convey how words might evolve to describe other varieties of *interception of air → [at the] throat* that are not *judicial relations*: lynching, murder, autoerotic asphyxiation, word-guessing games, rides in convertibles with scarves, Halloween pranks gone wrong. Complications aside, one can see the appeal of Wilkins's plan: not only does each thing have its place in the world, but each word conveys exact coordinates. Gone is the need for illustrative flowcharts. Gone are miscommunications and visits to the couples therapist.

For good reason, scholars have been obsessed with taxonomy since the beginning. How *do* we distinguish the pathological from the normal, the chemistry from the alchemy, the dick from the Dick? When *have* we deciphered everything in the world, or at least accrued the fewest casualties in the inevitable carve-up?

One of anthropology's founders, Marcel Mauss, aiming to better understand cultural difference, beseeched readers "to draw up the largest possible catalogue of categories.... It will be clear that there have been and still are dead or pale or foreign moons in the firmament of reason."[2] Rhetorical curlicues notwithstanding, his point stands: shards of disused or disabused logics lodged in grammar threaten the fragile bubble of meaning in generative ways. Proliferating categories has/have a purpose.

Wilkins's universal language project would have gone cold if not for an essay by Jorge Luis Borges published in 1952.[3] The particular citational route it has taken illustrates a larger point about the propensity of categories to congeal into stereotypes. Social theorists feel affection for Borges's mention, in "John Wilkins' Analytic Language," of a fascinating and possibly apocryphal Chinese dictionary and its incongruous taxonomy of animals (see Daniel's essay in this volume). But I've not read a single commentator who notes Borges's own somewhat ironic befuddlement, one he shows but doesn't tell.

To see it at all is, perhaps, to have been positioned askew to labels. Catching a glimmer of Borges's confusion offers an excruciating reminder that one's very sanity depends on being able to locate and never lose sight of the gaps among what is said, what is meant, and what one experiences – and then make some sanity-enabling peace with the resulting cognitive dissonance. Perhaps to share Borges's muddle is to absorb the full force of the exclusion at the core of assembling knowledge, whereas to *not* see it, or accede to it, is to avow and invite its confinement.

Unable to locate a copy of Wilkins's book (the subject of the essay), Borges relied heavily and avowedly on the Pankhurst monograph cited above. Nevertheless, his first paragraph introduces the problem of meaning: "All of us have once experienced those never ending discussions in which a dame, using lots of interjections and incoherences, swears to you that the word 'luna' is more (or less) expressive than the word 'moon.'"[4] Mauss might want to weigh in about the dead or pale luna. In either case, Borges offers his audience ("all of us") the lazy foil of a bored woman stuck at a cocktail party, while skipping over the truth that his essay relies on the genius of a "dame" – not to mention her sleuthing an actual copy of Wilkins's book. Without the insult, Borges could have made the intellectual point, yet the denigration is also part of the history of ideas, a story curbed and shrunken by its own embargoes. Who knows where the combined brilliance of Borges and Pankhurst may have led in a true collaboration? This example demonstrates the absolute significance of the slippage between word and thing, showing and telling, what is said and how it might be meant, and the difference between trusting categories and intercepting them.

In the meantime, scholars such as Pankhurst exhausted their own intellects on the make-work project of gaining suffrage – of becoming human. It took further decades of activism for public schools to desegregate, precisely because women and people of color were ensnared in pigeonholes, stereotypes that supported the physical, emotional, and intellectual dependence of many men on the labor of women and others in lower rungs of multifaceted social and economic hierarchies. Meanwhile, those drinking scotch in the smoking rooms of science, economics, politics, and tradition sat heavily on the lids of the category boxes. We finally admit that "boy" is an offensive address for men of all complexions, but the equally insulting "miss" for those identified as women but hailed as girls remains mind-bogglingly acceptable. Anyone who has been addressed in such a manner knows that, once calcified, such groupings take generations, money, imagination, activism, and energy to erode.

Drawing offers a new angle on the quandary of naming. Take a walk among the easels of a life-drawing session and note how differently individual artists represent the same body. Drawing, like fiction, is avowedly inventive, self and perspective injected through a specific hand and approach to line. Drawing animates, with charcoal or ink, a manifestation of our own internalized, socialized selves, expressed through fingers and eyes. To investigate how things are thinged, I hijack usual representational practices. William Kentridge said in a different context that when we put up a label, "we admit defeat," as the label "does the work for us."[5] Instead, I use the labels as a form of micro-resistance, adopting the master's tools (everything in place) to see what else can be built (shifting alliances). The label adds a new dimension for poetic exploration.

The blame for the shift in my drawings from a casual excavation of *cabinets* to a genuine interest in *curiosités* rests squarely on the menagerie "things that are not a pipe." On vacation in Paris, I had time to give over to my scribblings. Finalizing "things that transduce sound," I must have glimpsed one of the ubiquitous reproductions of *The Treachery of Images* (sometimes referred to simply as *Ceci n'est pas une pipe*, or *This is not a pipe*). My meandering pen seized the idea. For academics, citing famous predecessors is a kiss on the cheek from mummy; we revel in that warm fuzzy feeling of belonging to something grander than a lonely desk chair. So, this visual citation offered a comforting salve and also, as I found in writing this essay, a shelf of books dedicated to thoughts on Magritte's image, each jostling for a mention.

Magritte's painting offers a riddle: *What*, precisely, is not a/this pipe? The carefully scripted letters? The pigment? Oil? Canvas? It's certainly not the other translation of the French word "pipe": fellatio. (To be sure, the citational canon on that one is thin.) What did it mean in 1929 to paint an object-not-object on a backdrop of butter yellow, more akin in style to *Gray's Anatomy* than Vesalius's lurid landscapes? To be fair, the latter introduced labels to anatomical drawings, but Gray's bones and organs floated freely in the empty space of the page, predating Magritte's styled painting by some seven decades. Leave aside the Modigliani, Klee, and Miró that art historians tend to collate with Magritte. *Ceci n'est pas une pipe* belongs in the history of medicine.

Social critic Michel Foucault suggests that the painting is best described as a calligram – a poem shaped as the thing it deliberates, such as Guillaume Apollinaire's "Eiffel Tower" written in its eponymous shape.[6] In theory, words and things in the calligram, as in John Wilkins's language, become indistinguishable. A science drawing, on the other hand, purports to have two components: the thing and the label, each illustrating the other. *Spleen* labels spleen and vice versa: once you can recognize it, you can remove it. The authority of the science lies in part on the dual reference of saying and seeing for oneself.

The classification project depended on being able to represent the world in miniature – to perceive (a version of) a monarch butterfly or Mt. Kilimanjaro on a slip of paper, and to accept, from the garret of a dreary English manor, that such a creature or mountain exists despite its incomprehensible foreignness. The image/label simply *is* what it says it is. Thus, generations of schoolchildren, museum-goers, and pictorial dictionary readers have memorized and regurgitated all things great and small. Notably, the daylight hours we now spend studying were once used in physical learning, discovering the details of our local environments.

With this flourish, Magritte's knot loosens; indeed, the image/label technique – the very basis of Western knowledge systems – crumbles with the insertion of "not." Far from mutually illustrative, the picture and label are revealed through the sleight-of-paint *as the same thing*! And the curtain does not stop there. Magritte shows us not only that this painting is not a pipe, but that it wasn't a monarch butterfly flitting in the pages of your nature book, and no, you don't see Mt. Kilimanjaro on that postcard, not even close.

Under Magritte's measured pressure, the object/label system that anchors meaning itself has buckled, and more, the brilliant and regressive system equating label/representation/thing has been unveiled for what it always was – mere inscriptions on a mere page. Every child experiences a version of this disenchantment when, after repeating many times quickly the word "sweatshirt," a word that has taken a large proportion of their seven years on the planet to learn, they find nothing as warm and fuzzy as their favorite hoodie but merely a dry mouth-field of swishes and clicks: rtsweatshirtsweatshirtsweatshirtswea. Treacherous indeed, the label has *not* worked.

In truth, though, a pipe is a pipe and Magritte painted one (not a spleen or a robot). We all know it, recognize it, and agree. Reminding us of everyday suspensions of disbelief, he would not have become quite so famous (no matter how many friends and patrons supported him) if we didn't recognize the pipe emerging from the paint and cloth. Magritte's conceit reminds viewers that the maintenance of shared understandings about our things requires constant vigilance and care. A similar recognition lay behind universal language projects such as Wilkins's – a justifiable fear of absolute difference, indecipherability, and communicative deadlock.

Etymologically, "thing" derives from the term and concept of "assembly": we know a thing by the company it keeps, and the company things keep changes over time. For example, if Magritte ever thought about lungs it would not have been in relation to his pipe, but rather to tuberculosis, a main cause of death (with war and childbirth) for his generation. For us, smoking and lungs are intertwined. With the introduction of the cigarette, the pipe's disposable and self-immolating doppelgänger, everything changed. Those thin rods of plant matter and chemicals dominated the century in contradictory and largely invisible ways: at one moment the source of a buzz, an identity, or stock dividends; at another, the reason for a gathering to witness a relative suffocate to death.

The word *cigarette* barely supports the vast network of hope and trust, evil and banality in which the palm-sized box on the corner store shelf exists. (No word could.) A continual power play underlying the word and concept *cigarette* daily ushers it into being as a thing (the pesticides, pickers, factory workers) with meaning (the Marlboro Man, the surgeon general's warning) and effects (chemotherapy drip, teeth whitener, Duke University). With all of that, and the hyper-designed label on the box, the little punch of nicotine, the lurid smell of exhaled smoke, we create the sound, shape, and heft of *cigarette*. Today, smoking is different than it was in 1929 because of conscious efforts from all sides to change its meaning. And so the pipe is different, too.

The above list of details constructs a cigarette's cigaretteness just as surely as a cigarette is a pipe and a pipe is an object made of a small wooden bowl with a hollow stem. My own interests took my drawing toward smoking, but other artist-scholars may come up with other not-pipes to house Magritte's painting: things that are (not) famous; things that question the nature of a pipe; things (not) made in 1929, by Belgians, by (not) men; things that puzzle; things that may cause cancer or reproductive harm; things that have (not yet) caused one's own death; things known by educated global citizens; things that have too much written about them.

Eighteenth-century philosopher Denis Diderot intuitively understood the power of objects-in-their-proper-place when his campy new scarlet robe suddenly made the rest of his home seem shabby, with "no more coordination, no more unity, no more beauty."[7] The new robe sowed discord, when previously the "old robe was one with the other rags that surrounded [it]." The dressing gown spurred a "mad desire" in him to replace everything he owned with "new beautiful things" so that things matched, creating a collection that reflected properly on him (and his improved economic circumstance).

This epidemic of household consumption known as the "Diderot effect" is mimicked in all kinds of ways, from the number of words we use to describe our morning coffee to the "things some people jolly well eat but we tend not to"; the project of self-becoming through things and language involves judgeable judgment. But not just that.

Describing both the old and the new set of objects as beautiful, Diderot locates – disingenuously – the aesthetic offense in the *scrambling* of the two systems. Unlike a collage or flea market, a group of things becomes a collection by virtue of an organizing principle, a master narrative imposed as though it instead emerged through the objects within. This endeavor requires two things. First, a person who exudes vision: a curator rather than a hoarder. And second, a framework through which to reimagine objects that have been stripped of their native history and left stark naked and vulnerable to reinterpretation.

Theorist Susan Stewart notes that the whole point of a collection is to forget, to create an "infinite reverie."[8] Noah's Ark, an example referenced by both Stewart and our old comrade Wilkins, offers an example of the mechanism at play: each animal severed from their habitat attained a new, utterly foreign status as exemplary type, core progenitor, and DNA bank within the Ark's collection. Granted, a life-threatening calamity spurred this maritime trek. Nevertheless, for animals plucked from their favorite mud patch, their cozy gay lover, or an anticipated bramble of ripe berries, any captivity, no matter the reason, would be dismaying.

Collectors, entranced with their shiny baubles, have every incentive to remain blind to the erstwhile delights and sorrows of their new toys. Like the hunter with a knee on an elephant's neck, the theorist with a fountain pen, or the natural philosopher gauging the legitimacy of an egg-laying mammal, the collector draws on the aura of the collected to assemble a new, perhaps more expedient account.

A piece by artist Fred Wilson reminds his audience of the violence inherent in the nostalgia and disavowal of such daydreams. In a single glass display case, he placed a pair of blackened slave shackles and a highly polished, ornately wrought silver tea set. The title, "Metalwork 1793–1880," reminds the viewer of the semiotic work that made both objects possible. Unsettling the dust on the stuff stuffing stuffy history through the process of differently arranging things can provoke an uncanny recognition, an "aha" at once discombobulating and revelatory.

Things That Art, the graphic menagerie before you, aims to provoke new kinds of wonder at fragile, descriptive, predictive, contradictory, and unstable categories. It's an invitation to hug your inner platypus, kiss a hippo, and ignite the beam of your dame's luna.

Notes

1 Estelle Sylvia Pankhurst, *Delphos: The Future of International Language*, classic reprint (London: Kegan Paul, 1927).

2 Marcel Mauss, *Sociology and Psychology: Essays*, trans. Ben Brewster (London: Routledge and Kegan Paul, 1979).

3 Jorge Luis Borges, "John Wilkins' Analytical Language," in *The Total Library: Non-Fiction 1922–1986*, ed. Eliot Weinberger, trans. Esther Allen, Suzanne Jill Levine, and Eliot Weinberger (London: Penguin Books, 2001), 229–32.

4 Ibid.

5 William Kentridge, *Six Drawing Lessons*, Charles Eliot Norton Lectures (Cambridge, MA: Harvard University Press, 2014), 80.

6 Michel Foucault and Rene Magritte, *Ceci n'est pas une pipe*, nouv. ed. (Montpellier: Fata Morgana, 2010).

7 Denis Diderot, *Regrets on Parting with My Old Dressing Gown*, trans. Kate Tunstall and Katie Scott, *Oxford Art Journal* 39, no. 2 (2016): 175–84, https://doi.org/10.1093/oxartj/kcw015.

8 Susan Stewart, *On Longing: Narratives of the Miniature, the Gigantic, the Souvenir, the Collection*, 1st paperback ed. (Durham, NC: Duke University Press, 1993).

Contributors

Elizabeth Bradfield is the author of the poetry collections *Once Removed*, *Approaching Ice*, and *Interpretive Work*, and the hybrid photo-exploration *Toward Antarctica*. Her poems and essays have appeared in *The New Yorker*, *West Branch*, *Poetry*, *The Atlantic*, *Orion*, and many anthologies. Bradfield has been awarded a Stegner Fellowship, a Bread Loaf scholarship, and the Audre Lorde Prize, and her second book was a finalist for the James Laughlin Award from the Academy of American Poets. Founder and editor-in-chief of Broadsided Press, she lives on Cape Cod, works as a naturalist locally as well as on expedition ships in the high latitudes, and teaches creative writing at Brandeis University. See www.ebradfield.com.

Drew Daniel is an academic and a musician. As one half of the electronic group Matmos with his husband, M.C. Schmidt, he has released many groundbreaking recordings of electronic music fashioned from highly unusual sound sources, including washing machines, plastic surgery, and amplified crayfish nerve tissue. Matmos is known for its collaborations with a broad array of artists, including the Kronos Quartet, Terry Riley, Bjork, Young Jean Lee, John Cameron Mitchell, Robert Wilson, Anohni, and many others. In his other life, Daniel is a literary academic. He is the author of two books: *20 Jazz Funk Greats* and *The Melancholy Assemblage: Affect and Epistemology in the English Renaissance*. He is an associate professor in the Department of English at Johns Hopkins University, where he teaches courses on early modern literature, psychoanalysis, and literary theory. He lives in Baltimore.

Dr. Maria Dolores McVarish is an award-winning author and architect. Her forthcoming book focuses on the spatial history of race and landscape narrativity, coming to terms with little-known and largely disregarded figures in the American West. She holds a PhD from Stanford University in modern thought and literature. McVarish's architectural and design projects have been featured in *California Home and Design*, *San Francisco Magazine*, *Southface Journal*, and CNN's television series *Earth-Wise*. Her essays, drawings, and sculptures have been published in *Memory Connection*, *Diacritics*, *Zyzzyva*, *HOW(ever)*, *Architecture California: The Journal of the American Institute of Architects*, *The Art of Description: Writings on the Cantor Collections*, and various book collections. She is a senior adjunct professor at California College of the Arts, where she teaches visual and critical studies and interdisciplinary design.

appreciations

to kahlo and asha,
your birthright has been to forge
a new way through the uncomfortable categories
this book pokes fun at. it's not always
an amusing mission. you are invincible
and irreverent in the face of a world
that too readily reveals itself.

love and gratitude to dillon castleton, michelle murphy, nina wakeford, morlee griswold, abou farman, sherine hamdy, donna haraway, rebecca hill, yelena gluzman, julie livingston, blakey vermuele, david serlin, brian selznick, oxidate working group, kate zaloom, megan adams, rachael healy, jude feldman, heather love, bays, the write knights (rosie sims, nickolas suraway stepny, lienke diedericks), barb voss, lisa brown, christine byl, johanna drucker, lucy kimbell, barry blitt, nick sousanis, cori hayden, joe dumit, mario vella, cristiana giordano, and all my beloved chums and numerous friend crushes.

and an extra plump and appreciative drip of the pen to: maria m'varish, frances wallace, anne brackenbury, elizabeth bradfield, miriam ticktin, krista denio, jackie orr, derek simons, lisa feder, jake kosek, katrina karkazis, terry castle, drew daniel, joseph masco, julie ann yuen, diane nelson, frances phillips, asha marion, kahlo marion, samara marion, evelyn jain, anita jain, kamini jain, sudhir jain.

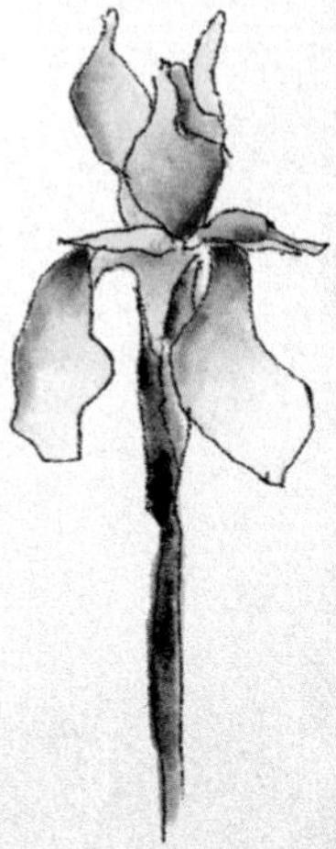

This groundbreaking series realizes ethnographic and anthropological research in graphic form. The series speaks to a growing interest in comics as a powerful communicative medium and to the desire for a more creative and public anthropology that engages with contemporary issues. Books in the series are scholarship-informed works that combine text and image in ways that are conceptually sophisticated yet accessible to broader audiences, open-ended, and aesthetically rich, to encourage conversations that build greater cross-cultural understanding.

Series Editors: Sherine Hamdy (University of California, Irvine) and Marc Parenteau (comics artist). **Series Advisory Board:** Juliet McMullin (University of California, Riverside), Stacy Pigg (Simon Fraser University), Nick Sousanis (San Francisco State University), and Fiona Smyth (OCAD University).

Other Titles in the Series

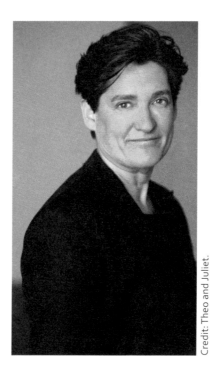

Credit: Theo and Juliet.

Lochlann Jain is a professor of anthropology at Stanford University and in the Department of Global Health and Social Medicine at King's College London, and has studied art at the Slade (University College London) and the San Francisco Art Institute. In art and in scholarship, Jain aims to disrupt common-sensical ways of knowing. Jain is the author of *Injury* (2006) and *Malignant: How Cancer Becomes Us* (2013), which won numerous prizes in anthropology and medical journalism, including the Staley Prize, the June Roth Memorial Award, the Fleck Prize, the Edelstein Prize, the Victor Turner Prize, and the Diana Forsythe Prize. *Malignant* was praised as "a remarkable achievement" (*TLS*), "a whip-smart read" (*Discover Magazine*), "brilliant and disturbing" (*Nature Magazine*), and having "the phenomenological nuance of James Joyce" (*Medical Humanities Review*). Jain's work has been supported by Stanford's Center for Advanced Study in the Behavioral Sciences, a National Endowment for the Humanities fellowship, and the National Humanities Center.